Meaning of Light and Life

Vibrational Matter Series Book 4

By Steve Preston

2nd Edition

2018

Table of Contents

4

Introduction

I guess you are wondering why anyone would put information about life and light together in a single book. The answer lies in very ancient documentation about a special type of light that generates life and/or allows an entity to travel from this universe to another. It sounds like a physics issue so I am trying to establish a broader description of light that fits better than what you are used to. Let me start this book with one of my favorite quotes by Sir Oliver Lodge.

"A fish probably cannot understand the existence of water; he is too deeply immersed in it."

This being immersed in light is one of the main reasons we cannot adequately define it. Like a fish accepting water as being there, we believe light is just there. What we are going to attempt in this book is to understand the water around us. Our 'water' is made up of 2 things; light and life. Some might think that our first topic is an easy one. "What is Light?" but I think you will soon see that light is not so simple. Unfortunately, we still need to address a number of things typically ignored in our school discussions about just about everything. Things like vibrational resonance of light and life, that 10 or 12-dimensional universe I have been addressing all along, lateral viewing of time, the concept of our neighboring-linked universe going backwards in time, concept of symmetry-of-energy instead of conservation-of-energy, and other items that simply must be substituted. If we don't do some redefining, anomaly after anomaly comes up in our efforts to understand things. I'm not trying to get you confused in this book. It is my last book on this subject and I want you to start expanding your concepts. Don't believe Peter didn't walk on water because someone said it's impossible. Don't think that you can't leave your body simply because it seems odd. Don't think that when Jesus told his people that faith as small as a grain of mustard-seed is all that is required to move mountains was a lie. Today new laws of physics tell us what Jesus told his followers 2000 years ago was perfectly ok in our universe; we simply have to be retrained a little. Jesus is God incarnate so it is not strange that he knew about these things before we did with our tiny minds. In this version of physics, called Participatory Anthropics or Quantum Mechanics reality takes on a submissive role and is only real when a cognizant observer is introduced. I know it is odd, but this new science gives us great power to modify our reality with something Jesus called "faith". Today, we call it cognition.

Once we get past the hurdles of loosening your ideas some more and trying to establish a semblance of definition about what LIGHT really is, we will go on to a subject that has plagued mankind. "What is Life?" Both have a similarity that will include some rather unusual descriptions of what you thought you knew. In a way, light makes the universe. In another way Life creates our universe. Both must be considered. While life must be indelibly tied the creator of

7

life; we will need to understand nature, cause, effect, care, understanding, and temper of God to describe a lot about our life's happiness, hurt, awareness of our existence. Finally, we will venture into the mystery of death. Surprisingly, there is a vast amount of knowledge concerning death that is both helpful to us while we are alive. This information, potentially, will allow us to almost revel in how death can help us. Let me start by stating that light is way more than simply vibrating electro-magnetic waves that sometimes have mass and sometimes do not as weirdo professors tried to tell you without chuckling.

Light and Life

This book is not a complex discussion that takes a scientist to understand what in the world I'm talking about. Mathematics and equations are not integrated into this book like the weave of a quilt and simple truths will become obvious with very minor equations and common sense. OK! It will get somewhat bizarre just like the others in this series, but I think it will open your eyes to a new possibility. This book doesn't stand alone. Instead, it is a companion book to the "Vibrational Matter" series. Hopefully, by looking a wide assortment of ideas and bits of evidence it clears up many of the anomalous details presented by other theories of matter. For instance, we now know that the atomic cloud defined by John Dalton in 1808 is not the essence of matter. There certainly are atoms or at least atomic characteristics that make up things, but basing matter on atoms and molecules and nuclear forces and covalent bonds and all the other things presented to you previously will not allow you to understand more basic elements of light or even life for that matter. All matter should be thought of as being made up of vibrations. I know you are familiar with the word vibration by now, but let me, again, clarify what I am saying here. I am saying that there is a nothingness that is vibrating to make mass and the same vibrating nothingness makes light. To make the statement even more bizarre, I'm not talking about vibrations moving through time. I'm talking about something that can be addressed within a time domain or be characterized outside of time so don't go sitting there and think that when someone says vibration that the little picture of a guitar string vibrating is the definition of everything in the universe. Vibration through time is only one emanation of particles and light. If you wanted to put a simple definition, right now, before we even get into the story, the only difference in matter and light is that they are going in opposite directions through time. Generally, this vibrating has no direction so it is sort of like pulsating. Just like the heart, the pulsating vibrational characteristic of the universe allows the universe to live. I know you are already wondering why you picked up the book in the first place, but I promise it will become cleared and your perception of everything will become clearer as you learn more about this thing we call light.

The fundamental characteristic of a pulsating universe, as we have been developing throughout the series, should change how you look the entire universe because we are not talking about something with height, length or width. Instead it is a field that vibrates and as the vibration's cross paths, they set up standing waves or nodes which we interpret as matter and light and everything else. Because matter is made up of this and not the three previously known dimensions, we must change our concept completely, but that is a different story that the first three books went over. I need to concentrate on LIGHT in this first section of this book.

Oops!

10

I know I said I was going to concentrate on Light, but let me very quickly tell you some details as sort of a review before we get started. First, we must understand that "String Science" assures us that at least 10-dimensions must exist to make our universe and to allow for that Big Bang thing that happened so long ago. If you have read the earlier books, you now know that we can also consider a subset of the dimension of time so we have 3-dimensions of particles, 3-dimensions of forces, 3-dimensions of life, and possibly, 3-dimensions of time itself. These 12-dimensions must work together and they all are associated with Light and vibrating nothingness. I developed a possible description of this new universe by building definitions of the following 12-dimensions that are required for our universe to exist and I think we should review them quickly to ensure that we are all on the same page. What we will find out soon is the following:

The universe not only exists to allow light, the universe exists because of light.

12-Dimensional Universe

Matter is made up of three dimensions of vibrating nothingness called Gravity, Aether, and Aetherio-Gravitation force also known as Nuclear Force. All energy in the universe is made up of vibrating nothingness called Magnetism, Electricity, and Electro-Magnetic Force also known as Phonic Force. Life is made up of three dimensions of vibrating nothingness known as Self, Soul, and Spirit-force also known as the light. Time appears to be made up of vibrating nothingness and is defined as forward time, backward time, and Forward-Backward Time-force also

11

known as space. Everything we have ever seen, thought about, felt, heard, witnessed, defined, believed, and hoped for have always been characterized as 3-mutually-perpendicular components working together but completely independent.

- The height, length, and width we all love are mutually perpendicular, work independently and are all necessary for something to exist.

- The God-head trinity can be defined the same way as the Creator-soul, the word-Self, and the spirit force/ holy spirit. What I mean by this is that everything that God made was made in accordance with his eternal being.

- Electro-magnetics has always been defined as three mutually perpendicular dimensions. [Electricity, Magnetics, Electro-magnetic force] and all other groups of three dimensions vibrate in a mutually perpendicular way so they don't disturb the others.

- Mutually perpendicular angles in mathematics must come in threes.

The first step in understanding light is to understand these basic building blocks or dimensions. The second step is to recognize that in a vibrational universe there really are no height, length, and depth dimensions making up the volume to put things in. Distance is a very arbitrary concept requiring our velocity to be constant and time to be an absolute. We now know neither is correct so we had to throw away length, height, and width. We have to change our concepts now that we are going to attempt to define light and the definitions must include this vibration thing.

Please throw away your old concept of light as these electro-

12

magnetic waves that come out of a flashlight. If you have to, simply take the batteries out of the flashlight until you can pull yourself away from that EXTREMELY limiting viewpoint.

Remember; we need to define matter, life, and energy in a vibrational sense or the universe is filled with so many anomalies that one could truly go crazy. Some of the more basic anomalies to length, height, and width are:

- **Sometimes light is a particle**, then, all of a sudden it looks like a massless wave.

- **Gluons**, Gravitons and other fermionic semi-particles don't exhibit length height or width, but they still seem to be real.

- **Quantum mechanics** is defined without any limitation of distance.

- **Scientists found red shifting** things in the universe that cannot be defined with distance.

- **Time travel gets all twisted up** if one tries to hold onto the "normal" dimensions.

- **Einstein's relativity** won't work with length, height, and width, because everything goes to infinity if one travels at the speed of light.

- On and on and on we could go.

Everything we are seeing today is broken down if length, height, and width are dimensions. They cannot be!!! Get them out of you head RIGHT NOW and let's get back to building particles a different way!!

Law of Entropy

This is where the LAW OF ENTROPY comes from. While people use the law to define things every day, but they don't know what the law of entropy really is. In this more defined universe. The Law of entropy is this.

Particles will decompose to the Aetheric "static" if they do not interact with elements of the operational dynamo. The Entropy Law simply says things will decompose to their lowest state, which has no real meaning in what we call reality, but we must understand the continuing struggle to make "reality"..

At the highest vibrational level, the Particle dynamo is characterized as a pure gravitational field or black hole. The highest vibration level of energy is pure magnetism. The highest vibrational level of life is pure spirit. Interestingly all three of these conditions allow transfer to our linked universe. If we talk about life, at very high vibrational levels we can control reality itself just like God Incarnate told us we could do 2 thousand years ago. In the Bible vibrational frequency is defined as level of Faith.

Cognizant Observer

The science of Participatory Anthropic says, if a cognizant observe does not witness an event in "reality" the event did not occur. The law could be expanded to say. Reality requires observation by a special type of life that has a Spirit. Live grass, for instant cannot affect our "reality" but you can. The little girl lifting a car off her father without he bones breaking or her tendons stripping away from the bone absolutely can happen. Spontaneous human combustion can absolutely happen. Anti-gravity can be real, and people can go to Heaven.

If you have ever heard about Schrodinger's cat who was locked away in a box with radiation that could kill it if it ate some is both alive and dead until some <u>Cognizant Observer</u> opens the box to view the event.

Scientists have measured the center of the universe point, but there is a problem. The Earth is at or near the center and, assuming the past existed before cognizant observers established reality, the earth would have been blown to pieces during the Big Bang.

Scientists wondered how carbon-based people could have come into existence at the exact time when larger molecules like carbon were being made and our sun had completed its initial states and now was boiling Hydrogen to make Carbon. and well before the heavy metals took over as dominant products which would have eliminated carbon-based animal construction on Earth. The answer, of course is we came into existence and because we were carbon based, the Sun's age was established along with everything else.

If we just try to understand the Biblical description of Faith we can see how things get here and how vibrating nothingness turns into photons and turns into light.

While I am at it, let me tell you one more oddness we get from the Bible. Genesis 1 indicates God made light 3 "ages" before he made the Sun.

Genesis 1:3-5 God said, - and there was light. —during the first Age.

*Genesis 1:16-19 God made the Sun. moon, and stars—
during the fourth Age.*

We could have discussion about how long an "age" is in the first chapter of Genesis of the Bible, but one would argue that without the sun, we would have no light on Earth, so we are talking about 2 types of light. One associated in some way with the sun and another that has absolutely nothing to do with the sun. In fact, neither is directly associated with emissions from the sun, even though we are taught that from an early age. The eye senses vibrating photons and sends the information about density and frequency as a chemical message to the brain. Most people convert those signals SOMEHOW, into light and nobody has told you, how. ---- It takes a cognizant observer of photons to build light in this reality. Once it exists, the other animals can use this manufactured 'light' as if they created it in this reality.

I get it; all this sounds like fantasy or that I have been smoking something but I hope you read the earlier books and you will stick with me here. I'm trying to slowly expand your awareness. There is a reason why there are so many ANOMALIES you either have to deal with or completely ignore when you try to hold on to what you were taught about length, width, and height being the building blocks of the universe. It really tries to tell you time is fixed, space is fixed, life ends when the body wears out and all your energy simply disappears. It also forces every religious person to separate Christianity from science so you have to live in an un bearable dichotomy.

- Incarnate God told us over and over that anyone could <u>move mountains</u> if they had faith; the old science said

16

reality was fixed.

- Jesus, as 100% man, made <u>water into wine</u> by looking at it; you were told reality is fixed
- Elisha made an almost <u>endless supply of Oil</u> from nothing and Jesus made enough food for 5000 at one time and 4000 at a different time from almost nothing; you were told that is impossible.
- Peter, Paul, Elijah, Elisha, Jesus, and many others brought <u>people back to life</u> and you were told its impossible.
- In the Bible, almost 100 separate common people <u>cured the sick</u> by touching them and again you were told it cannot be done.
- Jesus, Peter, and others <u>walked on water</u> and you were essentially told that was a lie.

On and on we see that it SEEMED TO BE impossible to have the Bible use the same physics as the Bible until someone finally changed the whole thing and now Religion and Science fit together, but you are going to have to use modern physics instead of what was taught to you; where a man can travel millions of miles and no time changes or matter changes in the blink of an eye into something completely different and time can go backwards.

While the book focuses on light and life, these are the truths that are most important to your day to day life. For instance, photonic light we normally think about as a component part of the what we call "Electro-magnetism" is not a true light. Light is established outside our brains or colors would be different for every person in the world. Red for one person would be created in the brain as green to another person unless it is a controlled element of reality. One characterization of light is something called "the Spirit". The ancient people tried to tell us about this concept thousands of years ago, but we would not listen. In ancient Jewish, Sumerian, Incan, and Biblical stories and texts, the spirit of being is described as "the light". After the Heaven Wars [The wars where Satan tried to take control of the place called Heaven] the punishment of the rebels was that their "Light" [or spirit] was taken from them according to a number of ancient works. Let me state this again. There is a special characterization we need to understand with this spirit thing as well, if we are to define light more completely. At the lowest level or static vibrational level, this Ethereal dynamo is characterized as "Carnal existence" such as would be expressed by a tree or something like that one could call "potential consciousness". At the highest vibrational level, the dynamo is characterized as a pure Spirit or life/field. According to ancient texts, it is this characterization that allows transfer to the adjacent universe that we could call heaven. Between the 2, all consciousness is introduced into our universe. The more carnal/ animal-like characteristics have similarity to the electrical component of the operational dynamo and the spiritual characteristics are similar in how they interact with the Magnetic and gravitational dimensional elements of the other 3-dimensional dynamos.

Sorry if that review was too confusing, but I simply wanted to reassert what you found in the previous books and see how Light and Life are affected before we discuss light in more detail.

Light Example

This is the same example I presented in the vibrational matter book, but it is just as revealing when trying to understand these new concepts. What do we know about LIGHT? If you said NOTHING! You would be almost completely correct.

Certainly, it is a particle or it couldn't exist. Right?

No, no, no my friend. Light sometimes is sort of like a particle and sometimes it acts like there is no-mass whatsoever, but there is an electro-magnetic wave or frequency associated with the color of the light; whatever that is! We know that the faster the "photon of light thing" vibrates, the more powerful it becomes. Soon, the fast vibrating photon thing becomes dangerous to humans as it can go right through the body [x-ray] and if its vibrating slows down too much it changes into something we call radio waves. I know you are thinking that these radio waves must not exist because they don't produce visible light and they have no mass, but let me assure you that sometimes these photon things do act like normal matter. If you look at the following diagram, there is a wiggly line. The faster wiggling represents a prime particle vibrating faster and faster. Radio waves turn into light that turns into the deadly gamma rays. The gamma rays would be like unto the deadly spiritual, magnetic, and gravitational elements of the other various

19

dynamos while the lower radio waves and slower components have a similarity to the Aether [mass], electricity, and the component we call life. It all makes a nice little bundle as everything in the universe seems to fit together in a nice bundle. Everything seems to be associated with other elements of the universe as it should.

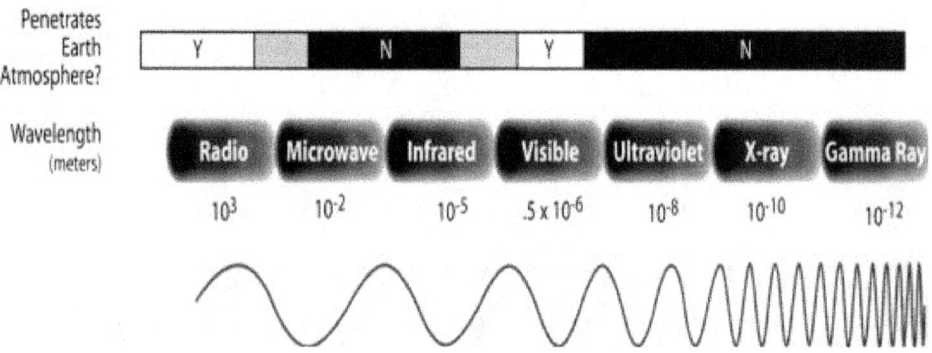

Like I told you every particle is actually a grouping of common vibrations. Sure enough, today the chart continues even farther and we will learn about that as we go along. With the new-found possibilities that all matter, life, and light could be defined as a vibration, sub-atomic theories started to become more and more splintered. Soon people didn't know what to think. Some of the previously crazy notions started to make a little more sense.

Sub-Atomic Particles

If you have wondered about anomalous characteristics like levitation, disappearing and reappearing, and even spontaneous human combustion, they are all supported by new developments and discoveries accomplished by these vibrating dimensions. Before we investigate details, let me say I'm sorry for the technical detail. It is not my intent on making equations fly around here so most of the data will be greatly generalized to make it simple. It is, however, important to open your mind to this concept. Without this backup, some of the details presented in this "liberation of concept" will not make sense and probably will be disregarded as too fanciful.

Sub-atomic particle research is a new science, which goes a long way in showing how levitation and most seemingly unexplainable phenomenon are possible. It also describes how photons miraculously appear, how atomic structure is determined and held, and how mass can be added and subtracted without time dilation. Everything we thought we knew has collapsed around us over the last few years and what has emerged is easier to understand because it doesn't have to have anomalies that could not be satisfied by the general knowledge".

Old Atomic Theory

In the past, we were taught atomic theory. In that basic theory, the atom was the smallest building block and all things are made from 115 different types of atoms. Scientists started to punch holes in the theory. They found Gluons, Bosons, Gravitons, Quarks, Photons and other particles much smaller than atoms. Here is the most fascinating part to me. Sometimes these particles just disappear. All the atomic theory was in jeopardy, but the theory continued to be taught.

Photons

Like I mentioned before, people started to look at the make-up of "light" and asked, "Where does a photon actually come from in atomic theory anyway?" What I was essentially told in college was that a photon was sometimes a particle and sometimes an electromagnetic wave. Just believe it and don't ask questions. The questions are only now starting to get answers and some of the answers make it look like levitation and even element conversion are both possible. This is neat! Maybe the ancient idea of changing lead into gold wasn't so wrong!

After all, the colleges are teaching that particles can miraculously convert themselves into electromagnetic waves. How much easier is it for a particle to simply change into another particle?

In school you may have been taught that if a photon's vibrational period is 1×10^{-6} seconds it has special properties called "Infrared". Even if you didn't take that class, the photons are infrared.

Of course, infrared isn't red at all. It's invisible to our eyes. It should have been called <u>Infra-invisible</u>. It was named a long time ago so we are stuck with it. The different colors of light miraculously mix together and somehow turn into white light even though you always thought that mixing colors together should make the colors darker and darker until everything was black. While you see white light every day, there is no such thing as white light.

White light is not really white light. It really is invisible light.

Therefore, "invisible" is simply the combination of a whole bunch of colors. As a single photon's vibration increases to a period of 4 $x10^{-7}$ seconds per cycles, it suddenly becomes "visible". As it speeds up even faster, the photon becomes an "X-ray" which can see through just about everything and then a "gamma ray" that can destroy tissue. The previous chart showed these things, but in this case, I want to bring out something strange. All this instantaneous changing and complete modification occurred in a tiny particle and it was accomplished [**without putting in huge amounts of Fusion energy**]. I know it sounds absurd, but people accept it every day without questioning. Here is my question to you.

"Just how does the photon particle change to another substance with entirely different properties?" We never answer this "simple" question. We just call it a photon.

On the following table are the common "frequencies" of photons that we accept without question.

23

Description	Cyclic period	Frequency	Feature
Helpful Infrared light	1×10^{-6}	30×10^{13}	Invisible thing
Visible light	4×10^{-7}	75×10^{13}	Visible thing
Dangerous X-rays	1×10^{-8}	30×10^{15}	Invisible thing that penetrates bone
Deadly Gamma Rays	1×10^{-9}	30×10^{16}	Invisible thing that destroys

You would think someone would stand up and say that is a lot of malarkey but almost no one does.

Photons and Gravity

Some new researchers tell us that a Photon can better be defined as nothing more than a Boson emitted from a particle-group that has absorbed energy.

Because of the energy boost, it must now eliminate the energy to insure stability-and here is the most important part. The most common energy absorbed is something we call gravity. As I mentioned previously, gravity is vibration just like everything else. The vibration level defines the [percentage of] gravity in a vibrational cluster we call a mass.

Don't get too bogged down in this. Hopefully, the last 3 books brought out most of this and I am simply expanding on the theme. If vibrational frequency of the electromagnetic wave is modulated by "gravity" the resulting perpendicular expression of this could sometimes take on characteristics of matter and sometimes, it would appear to be pure electromagnetic "unmodulated by gravity" waves. [This is the basis for the "sometimes a particle sometimes a wave"-description of a photon.]

Electromagnetic Modulation

Here is the basic element of the 10-dimensional universe and how we can or can't define light. By itself, electromagnetic dimensional components and Structural dynamo components react in only one way. The modulation causes stress in the universe or what we call "Force" is created. Forces are not recognized as a thing unless a "consciousness" is also integrated into the vibrational flow. It is this last characterization of "reality" that is the most difficult to understand and the most critical to understand in this book.

Hydrogen Turning into Helium

Here's an oddball question that I asked before. If hydrogen has one electron and one proton and helium has 2 electrons and 2 protons, can you put 2 hydrogen atoms together and make helium? The answer has been, "you can't do it without the exchange of a substantial amount of energy associated with "Atomic Fusion". It has something to do with what we learned in school that was called nuclear force [The force that "allowed" atoms to stay with large numbers of protons and electrons.]

According to the dictionary, a nuclear force is the "??" that is responsible for binding of protons and neutrons into atomic nuclei.

The dictionary doesn't have question marks, but when it comes down to it, they might as well have had question marks. It would seem that this nuclear force thing would pull all atomic structures to massive atomic clouds, so one could believe this nuclear force is a "reaction rather than an action". There are really 2 of these nuclear binding force things. The large one is the atomic bomb one and a very tiny sister is called GLUON. A gluon holds things called quarks together. If the quarks had not been held in place, there would be no need for the large nuclear force thing.

The Gluon Bomb

I would suppose if a nuclear bomb could level a city, a gluon bomb would level an atom. The electrons might have split and atomic nuclei would have been devastated. The scene would have been on the Monday night report to the other atoms and nuclear waste would have been all around. Luckily, to a gluon, some of these massive atoms are beyond huge, so these gluon bomb catastrophes are rare.

Aether and the Creation of Force

We now know that higher frequency vibration of Aether causes the appearance of larger atomic clouds and this is somehow associated with out-waves [produced by vibration nodes modulating vibrations away from the "nodes"] and in-waves [initiated outside our universe such that as the vibrations strike the nodes, stress is initiated in the form of what we call "FORCE"]. The last book in the series should have enhanced that insight and brought you closer to understanding the meaning of existence, but we must still go farther to the Ethereal Dynamo.

The Ethereal dimensional dynamo converts all this vibration stuff into matter, force, and conscious existence.

Ignoring this basic element, people create massive atom splitters to break the Nuclear force and make subatomic particles of all types. As the atoms are degenerated, there is a fear that the energy created during the separation of atomic particles could cause disaster or create a fusion disaster like a nuclear bomb or other scary things. One of these cyclotrons in Europe generates so much energy that there were concerns that they might create something called a black hole. In China, they are trying to be the first to develop and artificial SUN. If you think I'm crazy, you had better stay away from the guys trying to explain just what a black hole is and how they are going to contain a SUN. If one of these black hole things happened, the earth could explode and we would all be sucked into the next universe and if any survived the trip, who knows what we would find.

Don't even go there. You would not survive the trip without Armor and the armor requires something outside our basic concept of Carnal awareness, so don't go near a black hole right now!!!

Let's not think about these black holes and try to stick with things that are more manageable like nuclear force.

Matter and Light

Unfortunately or fortunately, nuclear force can be controlled. The force that holds atoms in quantized numbers of particles with quantized amounts of energy and density, apparently is a vibrational force. Here is one of my interests in this whole thing. Maybe we can get around the fusion reaction requirement that is so very dangerous to our existence with vibration. Light changes on the fly by external and internal excitations and vibrational resonances and if we learn more about how light does it, we should be able to do similar things to matter.

In a vibrationally controlled world, trying to modify this "apparent nuclear fusion" is not the only way to manipulate atoms. One can affect atoms by affecting the characteristic vibrations of the component particles or by affecting the apparent vibrational element of the entire atom {particle group}. These manipulations can and have been done without the huge explosion of a nuclear bomb. In fact, they may happen as part of a natural affect which emits photons.

By modifying particular frequency components of the atom, the various groups of sub-atomic particles apparently can and do become invisible, or will retrograde to another state, or may lose or gain their associated particle-mass energy by emission of that elusive photon thing.

That doesn't sound like light but it more closely defines how it is produced. I know that is a big statement so I'll get you more confused by defining invisibility. You can go back to just saying if this light "thing" vibrates faster its structure changes if you want. I don't care. The thing I want to bring up right now is that invisibility is somewhat different than you previously thought just like all other aspects of light are different than you previously thought.

Graviton and Photon

A graviton is a particle that has gravity and no mass. It's kind of the opposite of a black hole that has huge gravity and almost infinite mass Scientists see the effects all the time, but really defining what a graviton is can be a mystery. One way to look at the graviton thing is 2 fermion [Quasi-mass] things vibrating at the same frequency and in opposite phase. Everyone knows what happens. It's like noise canceling in those BOSE headphones. In this case the particle becomes invisible even though there are 2 of them just sitting there.

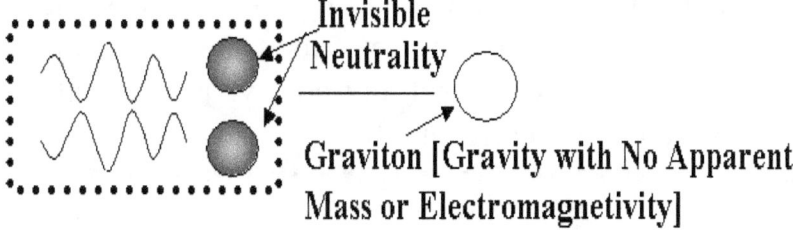

Noise Cancellation

Let's take a nice set of BOSE noise canceling headphones. These things work by bringing in background sound, inverting the sound and pushing the normal sound and the inverted sound into your ear 180 degrees out of phase just like the drawing above is showing 2 "quasi-particles" vibrating 180 degrees out of phase to one another. While you would think there would be more-sound for your ear to deal with, the background sound disappears just like the mass in a graviton. Don't even think this means that a graviton doesn't exist.

Light Similarity

Photons exhibit magnetic or electromagnetic characteristics and no mass. Guess what? A photon could be characterized by the same forward and backward quasi-particle that is synchronized in vibrational opposition and there you sense no mass at all, but a magnetic field is still generated. Others have a different opinion. Let's see what Walter Russel had to say.

29

Walter Russel

Walter Russel was a physicist that lived 1871 until 1963. His works are opening minds and shutting them at the same time. Some of his statements are provided below with some explanation. Don't take everything he says at face value, but don't ignore the basic concept either.

'Light is the living substance of Mind in action. It is the creating principle of the One substance. The One substance is the etheric "spiritual" substance of the One universal Mind. The entire "created" universe of all that is, ever has been, or ever will be, is but the One substance in motion, light."

So, you can see Walter tries to simplify the universe, God, and everything else by using light as the all-powerful creator. He also disavowed the tertiary dimensional dynamo that develops the spirit. Let's see if we can glean something that might be useful. By the way, the creator thing, should not be identified as the light. While Jesus indicated he was the light and the truth, and so much of the book of Genesis discusses light being given and taken away from individuals, one must consider that none of those things could be done by the creator, if Light was the creator.

Let's continue with more of Walter's insight. *"The material substance of Mind is an all-pervading ether which is indivisible, inseparable, indestructible, unalterable and unchangedable; but potentially it contains the appearance of*

all these dimensions of separability in the states of motion
which register the dynamic process of thinking.

OK! It sounds like he is sensing that matter must have electro-magnetic forces to exist, but later we find out that he throws away the concept of magnetism altogether. He also said that light is actually made up of two parts visible light and black light [non-radiating light]. This might make some sense. If light is going in one direction, the "black" inverse is traveling in the opposite direction to ensure that no physical change occurs in the universe. Unfortunately, he continued by excluding all components of the Operational dynamo except for electricity. Without magnetism, he constructed light as the magnetic characteristic in the operational dynamo that creates matter. He believed that there were 25 quasi-matter fermions and of them the quasi-component that established light was called Luminon. This was very close to what John Keely had called the Luminiferous Aether. We can assume Walter Russel was a student of the John Keely description of our universe. I'm not agreeing or disagreeing with the Luminon or Luminiferous Ether. I will say that "somehow" light and particles begin to be seen in our "normal" world when they vibrate correctly.

While there might be issues with his theories, let's look at this dual direction symmetric light concept. Here is the issue. *If light goes forward in time continuously, soon it will escape the universe and be lost.* Einstein had a horrible time with this seemingly destructive component of light. Milo Wolff more correctly identified an outside force of In-Waves that continually regenerated the vibrational components that continually leave our universe including vibrations of

31

LIGHT. Another way of describing the continuing entrance of in-waves into our universe could be this concept of black light or the reverse of light. No matter how one conceives the regeneration in our universe of light, ….

There is one absolute. Light must somehow be regenerated or it would, over time, be lost to the universe. If light is continually leaving our universe, SOMETHING is continually coming into our universe.

The Acceleration Field

If you remember, I explained how the dimensions that make up matter and the dimensions that make up electromagnetic waves act the same way in our universe. OK! They are backward to each other with respect to time, but other respects are similar. Here is an example of what I meant.

Assume that accelerating matter in space creates something we could call an 'acceleration field'. This is a calculation analogous to the 'electric acceleration field' that produces a force on an accelerated electric charge --- so we can assume that the 'matter acceleration field' in space produces a force on an accelerated mass m. In this situation, energy is transferred between the mass "fields" and electrical "fields". The force of the electric field and the force associated with the accelerated mass is an equal and opposite interaction. I know this sounds similar to Walter Russel's visible and black light characteristics, so don't discount that particular viewpoint if it helps you see what light is.

The opposing interactions are why our universe stays together and why light stays light.

We can either say that the acceleration changes the frequency of the matter waves by the Doppler effect or time is shifted to produce this Doppler affect. Both are the same thing.

Here is the important part. The resulting energy transfers to space and the accompanying **forces are localized**. They are the affect and affect local space-time while other areas in the universe may not be affected the same way as their localized characteristics are specific to that environment. I know this is weird, but here is the rub. These effects **appear almost instantaneously** [these are the opposing effects of all modification of substance in the universe] they are not delayed by space and time as testified to by astronomical observations and space missions. To make this discussion more to the discussion of light, if photons have no mass, where does the energy come from? If they do have mass, how in the world can they travel so fast? Maybe we had better start by finding out how light is made again.

How Is Light Made?

I know it's already confusing, but I need to bring all this into the discussions because we need to get rid of the belief that all matter is solid things and begin to appreciate this thing called vibration.

Let's look back at the Bible for a minute. All of you have read the first chapter of Genesis where the world was made, but I'm thinking you didn't really read it.

Genesis 1:1 The War

*Genesis 1:1 "In the beginning God created the heaven and the earth and the earth **became** without form and void"*

The book of "Jeremiah" tells us that the reason the earth got so bad was that Satan had the war and all the cities in the world were destroyed. Stay with me here. This will make sense in just a little bit.

Jeremiah 4:23-27

(A description of the end of the Heaven Wars) *"I beheld the Earth, and, lo, it was without form, and void; and the heavens, and they <u>had no light</u>. I beheld, and, lo, there was no man, and **all the cities thereof were broken down**."*

35

OK! At first it seems to be saying that all the stars had gone out, but, as far as we can determine, that did not happen. The word "Light" in this passage, like many others, is referring to the "SOUL". Without the soul there would be darkness. There would be no INTERPRETATION of light.

Genesis 1:2 No Light

Genesis 1:2 Then darkness was on the face of the deep and the spirit of Elohiym moved on the face of the waters.

Genesis 1:3 Light Before Light

Genesis 1:3- Then God said let there be light. God divided the light from the darkness.

Here is the interesting part of this verse. God made light before the Sun, moon and stars were shining in verse 16 much later in time.

***Genesis 1:16-19**-And God made two great lights: He made the stars also. And God set them in the firmament of the heaven to give light upon the earth, and the evening and the morning were the fourth day.*

Clearly this light thing described in the 2nd and third verse was not light from the sun, but it was something. I could get into my philosophy about what this first light was, but let me just say for his discussion that, apparently, God created three things

Three Creations of God

When I talk about his creations immediately you are thinking the universe, animals and man, but that is not what I'm talking about and it might not even be what he actually created. I think it is more esoteric that that. For instance, this whole vibration thing we have been discussing, I believe is the 1st of his marvelous creations. You are thinking "Vibration, anyone can do that!", but you would be totally wrong. No one knows what the vibration of creation actually is. It is so marvelous that probably the only one that can perceive it is God himself. The other 2 mysteries without reasonable understanding are light and life. Even though light is unexplainable, I'm going to do my best to bring out the nuances that make it God's second creation. In fact, all of these creations are intimately connected.

Vibration-The first was associated with the Structural Dynamo as he zapped into existence these "fermion vibrational packets". The result was that the physical heaven and earth was described.

Light-The next creation was in the Operational Dynamo as he created this light thing. Somewhat different that the vibrating fermions, "light" is something that is special that it extends beyond the limits of time. [Forget I said that. I'll get into that in a little bit.].

Life-Finally he created something in the Ethereal Dynamo we call "life". I don't mean he made DNA. There is one thing that is known. Life is not locked in the DNA. It somehow envelops it but don't be fooled into thinking we can create life by building a DNA structure or manipulating it. If life is already in it, we can modify that life only. The second part of this book deals specifically with this special dimensional characteristic.

Was Adam Created?

I know you have been taught that God made Adam and, according to the 2nd chapter of Genesis, he forgot to make

37

woman. Because of his mistake, Adam was asked to see if any of the other animals could be his helpmeet, but none worked out so God yanked out a rib and cloned him a woman. I believe the whole Adam thing is identified separately from the creation of the other living things for one reason. It had nothing to do with creation of a body, brain, skin, bones, and eyeballs. The difference in Adam's character and the other living things was that God reintroduced "Light" into one of his vibrational beings. The rib description was, most likely, a metaphor [I mean simile] as the spirit of a person does not live in the rib bone. Let me explain what I mean about the rib comment.

Rib and Light

The Hebrew word in Genesis that people interpret as rib is "*tsela*". The word is interpreted as "side" or "half of" everywhere else. It is odd to think that Adam had only 2 rib bones, so something is a-miss. I don't want to get into Biblical interpretation here, but one could imagine that half of this "light" thing that God introduced into Adam to make him "in the image of God", might have been used to make another human- "in the image of God". The Rib [tsela/half of man] and the light might have been the same thing.

Simple Definitions of Light

All three of these things [Vibration, Light, and life] are associated with light, but I will be concentrating on the middle creation for this definition. The descriptions won't necessarily make you think of ribs, but I want you to see that Light is more than beams from a light bulb. I think we should see where light comes from in our normal non-vibrating world. Simple---right? Heating up wood makes fire and voila! –Light-. Well, let's look a little closer.

Lasers and Fire

As I mentioned before, the way lasers produce light is simple. Electromagnetic energy is pumped into atomic clouds which push electrons farther from the nucleus than they want to be. As soon as this energy source is removed, the electrons zoom back into their original spin location and a photon is born as energy is released. Fire is a similar reaction except the electromagnetic energy is heat which makes electrons twist around to join with oxygen really, really fast. When the two atoms join, the electrons settle into a new path and photons are released. **What we see is this collapsing of the electron back from its excited state**.

There is an important nugget here. What this whole strange characterization really shows is that electro-magnetism and particles don't mix. [They are not in the same dimensional dynamos.]

Dynamo Craziness

I know you have been thinking I was crazy with all of this DYNAMO talk in this set of books, but there is an important nugget to get in your head. Particles of different dynamos don't easily mix. While Dynamos can come in contact and cause each other to react, the reactions must be neutralized quickly to be sustained in our universe or our universe can go crazy. The caution is that momentary joining changes the vibration of each. Electromagnetic waves change to vibrational patterns associated with light while the particle vibrational pattern momentarily increases. Its vibrational pattern during the encounter returns to its "Normal" vibration

39

as soon as possible. I know all this sounds like gibberish, but it's sort of a law that says that you cannot create or destroy matter.

What the law should say is "you can't permanently change matter without changing its vibration with an outside dimensional element.

In this case, the particles were affected by the electromagnetic dimensional elements to make light. We see these matter modifications every day and ignore them, because we are not looking at the whole 10 dimensions of our universe. Let's investigate these dimensions just a little more so we can understand what's going on around us. We need to reintroduce several things if we are to understand more about this universe and specifically light and life. You will see why I put these two things together as we go through this book. They are indelibly connected. Some of the things we will reintroduce in this book and expand are this concept of resonance, the concept of lateral time to define light, our adjacent 'backward in time" universe, this controlling element we call vibration and finally, some new ideas concerning the meaning of life itself and how we can control our carnal destiny and influence our spiritual one.

Space Resonance

I think I have you worried right now with the first description of how matter and electricity are correlated. Let me back up a little and restate resonance for this application in the words of Dr. Milo Wolff. **[my comments are in bold]**

"Resonance is composed of a spherical IN-wave which converges to the center [of the universe and comes from a different universe as a component of the operational dimension dynamo] and an OUT-wave which diverges from the center [of the universe and makes up what I call the structural dimensional dynamo]. Their separate amplitudes are [close to] infinite at the centers. [Like all other resonance factors in the universe, how close they are to being infinite can be considered the "quality of resonance".] When combined, the two waves form a standing-wave which has a finite amplitude at the center. The standing wave [appears] to be the structure of the electron. The inward and outward waves [sort-of] provide communication with other matter of the universe. Spin of the electron is a result of the reversal of the IN wave at the center to become the OUT wave."

While there is another definition I will get to later, this one helps us interpret how an adjacent universe "establishes" resonance in this world. The more we communicate with an

41

adjacent universe the higher in frequency our resonance becomes and its quality rises.

Quality of Resonance

Let me explain this "quality of resonance" a little. In electro-magnetics, quality of resonance describes the difference between the effect of a circuit outside its resonance frequency and that which can be described when it is in resonance. If a crystal is excited with a vibration that is half of the frequency it likes, it may vibrate a little and nothing more, but if it is hit with the vibration it likes, it begins to self-oscillate substantially.

It is this reaction that describes "quality of resonance". It is how well the device, or field responds to critical vibrations and ignores others.

In the electro-magnetic world, this quality depends on many things including what the crystal is attached to, how well the crystal is cut and how homogeneous the crystal is. In the electron or particle world, the same things can be surmised. Purity of the particle, and the things that surround the particle affect how close to infinity the standing wave appears.

Life Resonance

Now I'm going to let you in on a secret, but don't pass this on. Consciousness/life/ and death all act the same way. If we wish to affect the universe more and have a higher quality of resonance, we must become pure and surround ourselves with things that allow this pureness. OK! I can't exactly define what pureness is, but all this prayer and meditation stuff is probably more important to our quality of resonance [and our capability of affecting the universe] than one would

42

initially believe. I'm just going to leave it at that right now and get back to particles.

Resonance and Matter

I guess you are wondering why I even brought up this resonance in the first place, but **resonance holds matter together and it holds time together**. If enough electrons are in an area that are sensing similar in-waves, they align together just like a crystal. One could say that atoms are resonant plugs that are held together by like vibrations. Scientists found these things called gluons which seem to act in opposition to other particles and quasi-particles. Gluons hold quarks together and 3 quarks and an unknown number of gluons are called an electron. **Gluons are quasi-particles [fermions] that have a negative gravity.** That is; the farther the quarks move away from the gluons the STRONGER the gluon attraction becomes. It is sort of like the quarks are inside and invisible piece of matter that has a gravity. The closer they get to the surface of this invisible piece of matter, the more the gravity affects the quarks. The center of this gluon, matter would be the resonance point of what we call the electron whose resonance is defined by the vibrational characteristics of its component parts.

Gluons are not odd, they are simply invisible. One can say that they are this in-wave out-wave collision.

Who Cares About Resonance?

Why have I even brought up resonance? If a vibration node gets larger or smaller shouldn't matter to us. Right?

Well----we need to care for a number of reasons. Here are a few.

1. **The higher the resonance** of electro-magnetics, the closer it comes to being light which is the most stable electro-magnetic form and the most useful form of electro-magnetics to the universe.

2. **The higher the level of resonance** of a particle or quasi-particle, the more stable the matter it establishes becomes and matter is the most useful form of a quasi-particle to the universe.

3. **The higher a person's resonance is**, the more he or she can affect the universe and their own characteristic universe becomes more in sync with everything else. The closer one can get to God.

Resonance & Quantum Mechanics

I know you are wondering if I will ever give you a definition of light in this book, but I'm trying. You cannot simply blurt things out or they may not seem true. In order to allow for you acceptance, I need to present you with intermediate definition. Lots of definitions here as we look at the idea of quantum mechanics. Scientists, generally, have no idea what quantum mechanics is, but let me give you sort of a bird's eye view. It will allow us to understand light better.

Niels Bohr [1885-1962] tried his best to mess up our minds. He indicated that there is random vibration associated with everything, but if you test the spin of an object, and it is going one direction, there must be another that is indelibly linked and going the opposite direction. If that wasn't odd enough, he indicated that the linked objects could be millions of miles apart when the transposition occurs. Einstein indicated that the whole quantum mechanics theory was junk science that violated all of his theories. Einstein died before tests were devised that generally proved the phenomenon. --- Not only had the linking be seen, but also this elimination of time space as a boundary has been witnessed. Before he died, Niels, was writing something entitled "Light and Life Revisited" what he had finished was published after his death, he understood how very close light and life truly are and this quantum mechanics thing may hold a key to help us understand it as well. My version of light and life is a little different than Niels, and it has a lot less math, but the idea is similar. One cannot understand light without understanding life and vice-versa.

You may ask how does the impossible aberration of quantum mechanics and elimination of time-space go along with vibrations that absolutely need time-space and the idea of resonance as the builder of stability in the universe and as a way to remove the confines of ENTROPY. Well, there is no simple answer, but here is another bird's eye description that may tie things together and allow us to, at least, breathe in this universe that is making less and less sense, instead of becoming more and more defined.

Let's say photons or the electro-magnetic vibrations that can be characterized as photons are zooming through our universe from "somewhere else". If they enter our universe, a component of another dimensional element must be leaving at the same time or the universe would have no characterization at all. This is sometimes referred to as **conservation of existence**. OK! I just made it up, but that is what it should be called. It is this conservation of existence that actually is characterized as quanta as defined in the preceding overviews. As energy is introduced into a vibrational node to "Synchronize" an electron spin---- an equal and opposite removal of energy must be generated by the same electromagnetic influx or the dimension is not stable, symmetric, or defined. If we understand that nothing is "created" but simply changes state, the electron could not lose its "essence" [I want to say Aether] that allowed spin in

46

all the other directions unless some Aether is introduced in a linked electron that sort of pushes it the other way.

Eliminate Time and Space

I know you are saying--- "What in the world has he been saying?", but I think some of it will begin to make sense as you go through this series of books. If you are not saying that you are possibly saying, "What about the elimination of time and space in quantum mechanics?", so here is what I believe. In-waves that introduce force and out-waves that establish physical characteristics are each associated with "times" that are backward to each other. While this allows work to ensue, it does something very important besides that. It eliminates time during the exchange.

Simply put-- backward and forward time together make the absence of time and quantum mechanics can be more easily understood.

Because there is no time generated during the exchange, the exchange cannot be recorded by us. Even if the exchange took a thousand years, there would be no recognizable time for the action. Here is the scary part. Even if one were to go the speed of light there would be no recognizable time.

What?????!!

[Sorry for that last question mark and the 2 exclamation points, I was getting frustrated.] Here is the leap of faith. If there is NO TIME, there is no space. The actions would be resident at any or all locations at once. This, I believe, is where Einstein began nodding off. He was old and tired and unwilling to even consider the elimination of space. Neils Bohr was still young and thought about all of this weird stuff.

What this does is--- it allows us to transfer to anywhere anything----at least the essence of anything--- instantly just like the Star-trek "Beam me up Scotty" thing. The reason one could travel the distance is that there would be no distance so long as one could somehow introduce in-and out-waves together. That is where resonance comes in.

If someone wants to introduce a modification of space-time, all that is required is to temporarily change one's resonance.

Photons do it all the time and sometimes they are particles and sometimes [instantly and possibly simultaneously] they disappear and act only as waves.

Resonance in the Old Days

Jesus, his apostles, Moses, Elijah, and others from the Bible would have used this simple technique to turn water into wine or blood. Simply change one's resonance and something has to give to correct the shift. Do it right and things seem to be miraculous. Don't get me wrong about what I'm saying. **Jesus was God incarnate**, but while he was on Earth he was 100% human and he taught some to understand what he meant when he said "With faith of a grain of mustard seed **anyone** could move mountains." He could just as easily have said that changing your resonance by elevating your awareness to the non-carnal universal of heaven would allow one to control many carnal things simply by removing time-space during the transition for this equalization known as quantum. I guess his apostles looked

48

at him weird enough when the moving mountains comment was made, so he tried to keep it simple.

Resonance and Light

Light can do so many things like x-ray, and making its particle nature disappear and all the other things it can do because resonance is a powerful component in our universe. Later, as we discuss life, please understand that resonance is the method to modify the "apparent" vibrational characteristics of life and this can allow a completely different type of life if you want it.

Life Section Adder

I know this is in the wrong section, but there is a reason that positive thinking brings good things to us in life. The reason is that we affect our consciousness resonance. Resonance is one of the characterizations that ESTABLISH quantum mechanics and when it occurs we see miraculous things. Positive thinking goes way beyond the limited descriptions of books and commercials on TV. This whole concept of moving mountains by faith and walking on water are emanations of that understanding. Mankind, today, is trying to understand the concept by brute force and trying to remove the consciousness away from the equation and they are having limited success.

Today

Today a number of scientists are teleporting particles between distant sites using this quantum mechanics stuff. The particle essentially disintegrates at one site and is regenerated, instantly, at a remote site by characterizing the spin and essence of the original particle just like Neils Bohr had predicted. I know this is a somewhat different view of quanta than the simple globbing together of electrons in rings or channels, as we sense the atom to be and I understand that this does not give you a warm understanding about photons, but we need to, at least, recognize that when time-space is eliminated, there is only one mass to contend with in this universe. Individualization is ONLY characterized by consciousness rather than bodily differences and there is only one light. Let me say that again.

In a quantum mechanics defined universe, there is only one mass and one light. The only reason we accept differences is by means of our consciousnesses and its control over resonance.

Now that that is crystal clear, let's look again at lateral time.

Sorry for the "now that that is crystal clear" comment. I know this is all sounding way out there to you and if only a piece of it starts sounding possible, it will open a vast new view of our universe, so please stay with it.

Lateral Time

Quantum mechanics has gotten me somewhat confused just like Einstein had been. This whole "one-light" thing is modified in time as resonance is altered by outside characterization. We can potentially see the changes by simply viewing what I termed in the predecessor books as sideways or lateral time. A person viewing time laterally would see the beginning and end of time "at the same time" and he would see the progression of light as a time-based phenomenon. I know that sounds like the quantum mechanics thing again, but this is different in that it is simply viewing everything differently rather than initiating a causal event and seeing a resulting opposite event millions of miles from the first event. I put one of the lateral time diagrams on the following page as a review from previous discussions, but you can see that god can see the beginning and end of your life simultaneously and everyone is piled on top of one another spatially. The excursions of the graph represent light going forward and backward in lateral time's "equivalent of time". We could call this equivalence "mass resonance change" in our normal time perspective.

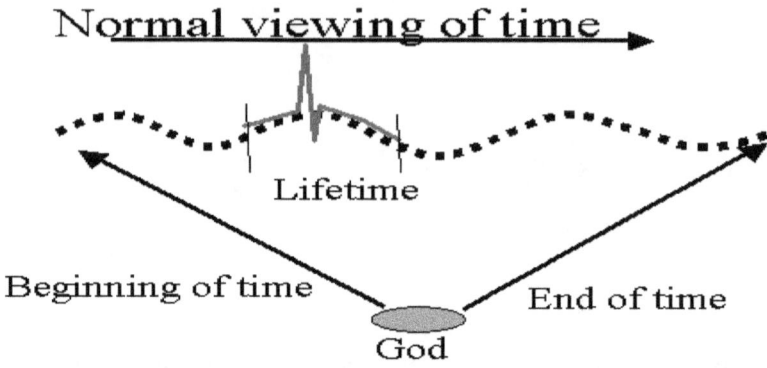

When God said he knew the beginning and the end, He would have had this lateral view of everything. He would see, for instance, your life all at once and in place of time, he would see "modifications of resonance". I know this is all new to you so I will try to explain it as I can so you will appreciate what light really is.

Now for a hard question, if everything is really made up of vibrations only, as I have presented in the first 3 books and this one, what would everything look like when viewed in **lateral time**? Hurry up; the clock is ticking. -----Come on!

-
-
-
-

If vibrations are emanations of modification over time, the answer would have to be a "solid mass"?

This solid mass isn't a mass at all. It is simply a compressed vibration. OK! I don't know what compressed vibration of nothing really is. It is easy to write about it and make you think I know something special, but it is quite another to be able to picture what light is in your head. It all has to do with perception. As we look at lateral time in more detail we will see a stronger relationship between life and light.

One can say that life is light viewed sideways and light is life viewed sideways.

I will get into lateral time view point more later as it is truly a wonderful way to understand how light and life are interrelated. Certainly, no one would suggest that a light bulb had life, but what one may find out is that there are a number of similarities between light and life and some critical differences. The main differences are in the Ethereal Dimensional dynamo. Before I get back into these more exotic descriptions again, let me back up a little and try to reintroduce light as defined in "normal-time". Then we will combine it more in the lateral time world and finally, I will add in the more difficult theme of "Life Definition".

With that, let's look at light more specifically as this stuff is making my head spin.

A Light Description

Look up at the sky. Light is everywhere but no one, I mean NO ONE, knows what light is.

I know it's something that allows you to see things as vibrating nothingness reflects off other vibrating nothings, but what I'm really talking about is the essence of light not what it does. Some tell you it's a particle that has no mass, others say it's something they call an electro-magnetic wave like a radio signal. Others get more scientific on you, but the result is the same. They tell you some of the characteristics of Light and leave most of them alone because they are simply too scary.

Make no mistake. LIGHT IS scary!!

I don't have any trouble bringing up scary things so this book will attempt the seemingly impossible task of defining light. I must admit to you that some of the information you will read will be odd to you because we cannot define light easily. Light, for instance doesn't always exist in this universe. When it does exist, it is responsible for many things. In fact, the vibration we use as a reference of light or electro-magnetic energy is one of the dimensional elements of this universe. It is one of the things that holds our universe together.

Let me ask a light question. What is the color of red light? While you think you know, No one can answer the question, like I said before. I know someone will say it is electromagnetic waves vibrating at 400 to 480 terahertz, but that is only a tiny part of the answer. Vibration only vibrates things. What makes it become "light" that opens our minds to the world around us? How can we see flashes of light and dreams with our eyes closed? How does light get so destructive simply by changing its resonance which modifies its frequency?

Who was that guy that said that light [or possibly visibility] was the opposite of absorption? If these terahertz things bounce off walls we see light. OK! It sounds stupid to me too, but that is what is typically said. That is what is being taught in schools today.

These vibrations hit cones and rods in our eyes. They get excited just like a crystal gets excited with some electric current is put across it. When that happens, guess what they do! Correct!! They vibrate. Do you suppose they give off light?----No!!! They cause chemicals to be generated, changed, transferred, and linked to the middle of the brain. Still there is no visibility and no red light. Hopefully, by now you understand that everything is vibrating. Is everything light?

After "some" vibrations cause chemical reactions in rods and cones of the eye, something magical happens. The brain [or your consciousness] somehow decides it can use the impulses to create a visible world and bring it to life. In our dreams, we do the same thing, but without the initial chemical change.

Stick light

If we try the experiment with a "vibrating stick" shoved in the eye, the rods and cone vibrate and send messages to the brain in a similar way, but the brain, for some reason, doesn't like the frequencies, so everything stays black.--- I caution you to do this stick light experiment slowly. I know the eye hurts, but we had to do the experiment to show that vibrating sticks and vibrating photons were different.

Radio Light

If you think that we should stick with electromagnetic waves, we can use radio waves hitting the rods and cones. They make the rods and cones vibrate and messages are sent to the brain. The brain will not open up a world and allow us to see the hue of radio. I know what you are thinking---The radio waves are going at the speed of light so that the rods and cones could not react with the waves because time would be stopped for the radio waves.

WHAT JUST A MINUTE!! Light is going the speed of light as well. How in the world can the rods and cones of the eye sense light in the first place????

The only answer I can come up with right now is that light is a figment of our imagination.

We'll do some work on this definition as we first try to understand light going the speed of light and faster than the speed of light. Then we will turn our attention to the easy question----What is life? While the question is easy, the answer is a little tricky just like defining light.

While we will see that light is, sort of, holding the universe together, sometimes it is used to simulate weapons and that is where I must describe what I do.

My Background

OK! I am shooting out all types of words and comments that sound like I'm just smashing syllables together or combining comments to make sentences long, but I truly am not trying to mystify. I'm trying to open up a new insight. As this investigation gets more intense, let me identify my level of insight in this matter, so to speak. By trade, besides writing, I am an electro-optics engineer and I design force-on-force training devices using lasers for the U.S. Army. Lasers are used to simulate bullets being fired at opponents instead of using real bullets, because they don't hurt as much. These packets of photons have certain issues that must be regarded if a successful system is to be produced and we can learn something about photons as we investigate these items. The things we need to consider are particulate attenuation, absorption, scintillation, and photonic density.

Photonic Density

If one is to use light to simulate a bullet, you must first build a photon gun that has enough energy to make it to the intended subject. The best way to do this is by using a laser transmitter. Not only can the light more easily be focused in the intended direction by one of these laser things, but also the wavelength of transmission can be controlled by the type of atomic clouds being excited. [Later we will look at lasers in more detail.] Knowing the wavelength allows the detection systems to be "tuned" to the transmission wavelength so that other photonic sources don't confuse detectors placed on the bodies of the intended targets. By using absorptive filters, only the intended wavelengths can be received and the target knows that the specific laser energy used in the simulation was received. What do we understand from this?

Light "apparently" can be easily reflected or absorbed by materials and we must understand this effect if we are to understand photons. How can light that has no mass be reflected in the first place?

Scintillation

As the vibrating laser beams are transmitted, they encounter air and particles. The particles in the air reflect or "absorb" the photons just like the special filter materials used at the targets, but there is another affect to consider. Uneven heating of air causes "small packets of similarly heated air globs" all over the place. This is called scintillation, the greater the difference in heating of the air, the greater number of these little globs. Different temperatures mean different pressures and air densities in these globs. Different pressures and densities affect photon's direction of travel. More pressure or density means more reduction in photonic speed. In a lens, when photons enter the glass, they slow down the same way. Several things happen, but the most noticeable is that the direction of light leaving a lens is different than it was when it entered. These little scintillated air globs do the same thing as a lens. Because there are thousands of these different sized and different shaped lenses, the light beam is moved all over the place. You see it all the time as wavy images across a desert or highway. The wavy image is caused by the scintillation globs moving photons differently. What we see from this is as follows:

59

Photons change direction and slow down as they go into more dense materials or higher pressures. It is known that the speed of light in a vacuum is faster than it is in our air. We must understand this feature as well. A question might be, "Why would no-mass photons slow down?" A second thing should be asked as well. If light is not traveling at the speed of light, what should we call it?

Possibly we could call the light we see in our atmosphere as "almost light". In the vast regions of space, where light goes faster, it could, then be called light so that Einstein's speed of light definition makes more sense.

All these things, absorption, attenuation, slowing down, and reflection of photons don't make sense in one way or another. We need to get a better handle on light and "almost light" or nothing will make sense to us. For our first step we need to investigate vibrations.

By the way, it is the slowing down of what we call light that actually allows us to use it to see with, so don't be afraid of almost light.

Here is what you need to understand right her and right now, because it also is important when trying to understand life and death.

When light, or anything else goes the "speed of light", it sort of ceases to exist. Mass must be stripped away before that point can be achieved, but Life, Light, Nuclear force all disappear in our "timed" world if they attain this "terminal velocity of the universe."

Light Vibrations

So someone told you that different colors of light were different vibrational frequencies of light and you believed them. Have you ever wondered about these innocent vibrations? To gain insight on light vibrating we must look at some things we "loosely" call atoms. I'm sorry for the confusion you will feel in the next few paragraphs, but it cannot be helped.

Atoms Aren't Atoms

I know you have been told all your life that atoms make up everything, but that is pretty much an incorrect thought. While the characterization of the atomic cloud does, in fact, produce things that have similar chemical structure, new science suggests that atoms themselves are made up of things that don't exist. Experimentation kept breaking down the components of atoms farther and farther until these quark things we defined as the particles that combine to create an electron were found to be made up of tinier things called fermions that don't really exist. One of the fermionic things is called a gluon which acts like glue to hold the quarks together. Scientists just "know" these gluon things are there or matter wouldn't exist.

Fermions Are Not The Answer

Frustration started setting in as these fermions started showing up. There are all types of fermions and one can even classify a photon as a fermion. I know I haven't redefined fermion for you yet, but it's coming up. Fermions are interesting things [or should I say quasi-things] because they are missing something that keeps them from becoming particles. The graviton, for instance, makes its own gravity, but it has no mass at all and that is an impossible thing. While it is impossible, these fermion things do the impossible every day and putting enough of these fermions together makes an electron or a proton mass or an atomic cloud.

Don't get comfortable yet. While the photonic fermion is a way of describing light, it does not define it.

We can think of these fermions as vibrational energy packets and it's the vibration word we need to concentrate on.

As I keep saying, everything seems to be a product of vibration in one way or another, so instead of describing things as mass particles, a better way of looking at everything is by its vibrational essence.

Einstein's Unified Particle Found

These new fermion things are close to the unified particle Einstein was looking for his whole life, but, unfortunately, there are all types of fermions, so we have to looks at what makes up fermions to find the elusive unified particle. The "people who know" tell us that fermionic differences all come from differences in their vibrational patterns. By changing the vibration, completely different particles can be created or changed.

Wow! We can surmise that the unified particle of matter isn't a particle at all. It is the vibration of nothing or almost nothing.

Now For The Confusing Part

If adjacent patterns have identical vibrations and are in phase, the fermion comes into existence. It miraculously obtains MASS. Get enough fermions and a boson or a quark appears. A bunch of quarks makes an electron and still more can "gather" to create an entire atomic cloud. The same things happen with the special fermion we call PHOTON.

Fermionic Photon

This almost impossible to perceive vibrational pattern we call a photon can be absorbed into substances to help create bosons and electrons and the whole bit, but when it is free, it sometimes acts like particle and sometimes like an electromagnetic wave, but one thing is for sure in free space, it has a forward motion that equals the speed of light so the essence of light is traveling faster than the speed of light. The diagram following illustrates the point.

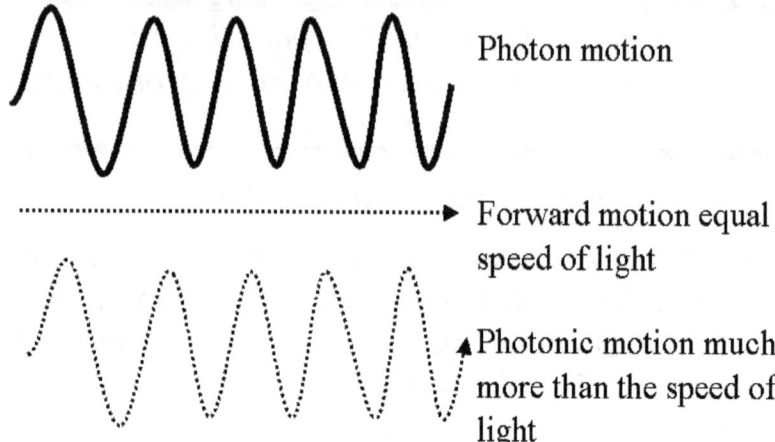

Photon motion

Forward motion equal speed of light

Photonic motion much more than the speed of light

Light is Faster Than Light

If the forward movement is the speed of light, the oscillations represent the travel experienced by the photon itself. That zig-zag, or backward and forward, or in and out motion assures the photon is going very, very fast. If you remember, I already told you that light doesn't exist at Exactly the speed of light, but we will see that faster is just fine.

With Einstein telling us that mass goes to infinity at the speed of light, we determine that faster than the speed of light is like going backwards in time. This backward time idea disturbed me so I thought I had better figure out more about this unusual thing we call light.

An Anomaly

Let's look at Einstein's old paradox or paradox generator. Two people are together and the same age. One goes traveling at close to the speed of light for 20 years. Neither shaves until the traveler returns. When he returns, the fast guy has not aged a day and the other guy is 20 years older and has a 20-year beard as shown below. It didn't matter where the traveler went and how he came back so long as he did it close to the speed of light. He could even stop every once in a while to see stuff. How can this be? If both are in the same universe, they must be affected by time in the same way. OK! I have already indicated that they aren't, but humor me right now.

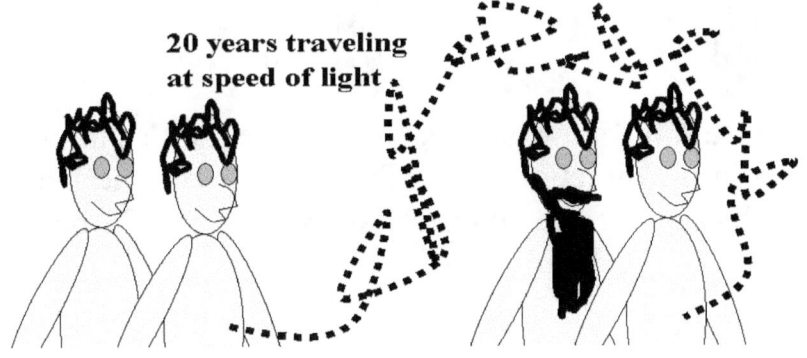

20 years traveling at speed of light

I know I have brought up this example previously, but this time, let's have the man going the speed of light, simply spin in a circle and see what happens. While he is spinning, his size goes to zero in the direction perpendicular to his speed and his mass goes to infinity [according to Einstein].

Assume that the now tiny, fast moving man held a flashlight outward and secondly, the spinning man was not just moving around in a circle but was spinning in all directions and you see that he becomes a speck of light. The next time you see light just think of it as a spinning man and time travel may become a little closer.

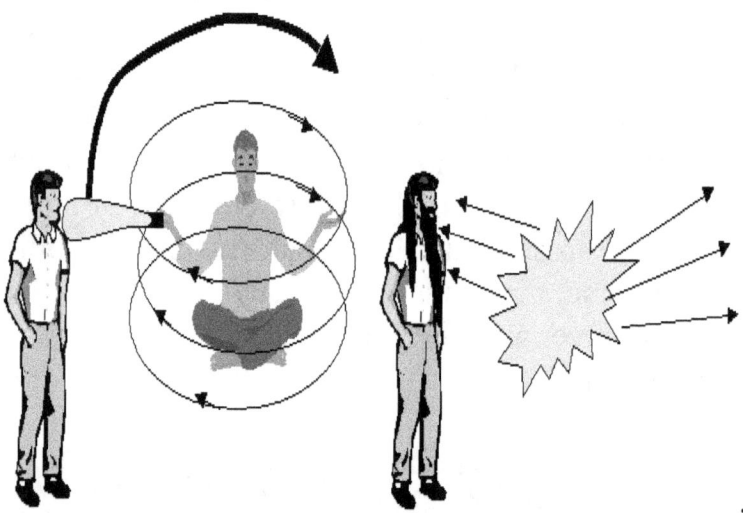

What would the slow guy see? Each time the flashlight came around in his direction he would be blasted in the face with light. Because of the spinner's speed, that would be all the time.

To the slow man, he would, sort of, see his friend turn into a light.

Tiny

As the fast man spins faster he gets smaller and heavier. Close to the speed of light, the spinning man is very small and very heavy. Here is the part that splits reality. Because the spinning guy stops aging, time would almost stop. To the slow guy, the vibrator actually traveled into the future. So, this whole-time thing is pretty volatile to our consciousness.

Another Anomaly

Let's take light, or photons, vibrating such that it makes an entire vibrational cycle in 400 nanometers as it whizzes around at the speed of light. The light hits your body and sort of bounces off of it. This isn't exactly what happens, but you get the idea. Light is reflected off your body. Now for some oddness. The photons get mad so they vibrate faster [about 1 x 10^{-12} meters per cycle.] This time when the light hits your body, it goes right through you and on the way, it destroys many of your cells. The beam of light is so happy. Unfortunately, you are not in good shape from this "gamma radiation" that you encountered. Why did light change and become so nasty? I know I have presented this in one way or another in the previous books, but ------

Remember this oddness. Light does this nastiness and it has no Mass [most of the time.]

Bigger isn't more powerful in our universe. The only thing that matters is how fast you can vibrate.

Difference Between Light and Photons

67

You may still believe that light and photons are the same thing because it has been pounded into you. [Oh boy!!]

Potentially, they can be associated with one another, but there is a major difference. Light, or more precisely, the speed of light is constant to you. I didn't say light was constant. It is constant to your "normally resonating consciousness". Einstein found this out, but he tried to keep light and photons together and this is where years and years of confusion have been billowing up.

Remember the flashlight shining example? If you shine a flashlight and you are going 1000 miles an hour, the photons will travel from that point at the speed of light [right?] Remember that your mass is slightly elongated and thinned to stay in line with the theory of relativity.

How Fast Are You Going?

If the 1000 miles an hour worries you, you are in for a shock. If you live in Florida, you are spinning in a circle about 1000 miles an hour already as the Earth rotates. Because the direction is across your body, you are fatter on the equator than at the North Pole where you would go slower. If that isn't enough, we are spinning around the sun at a rate of 66,000 miles per hour. Besides that, our solar system is milling around in the Milky Way at a rate of 43,000 miles per hour.

Now for the bigger numbers the Milky Way is rotating such that it takes 225 million earth years to make a galactic rotation. This means that the galaxy is turning at a rate of 483,000 miles per hour.

We certainly aren't done yet as the universe appears to be expanding. While expansion is an arbitrary term, red-shift analysis tells us the Milky Way Galaxy is moving at a speed of 1.3 million miles per hour. We are moving roughly in the direction on the sky that is defined by the constellations of Leo and Virgo.

That all being said you are moving at least 1.3 million miles an hour so how fast is light to you? 186 thousand miles per second is always the answer. Someone seeing the light that you are shining that is "stopped" in space sees a light shining at 186 thousand miles an hour as well, but you know in your heart it must be going another 1.3 million miles per hour over that 186,000 miles per hour. [It is not!] This is where light and photons separate. Electro-magnetics is constricted by the in-waves and out-waves of the universal sphere. But light stays constant to a conscious entity because consciousness is one of the dimensions of what we can call our "personal universe".

There is a change in the light, however. To the stationary observer, the number of cycles a photon vibrates while it is moving is less than experienced by the person on the earth moving so quickly. To the stationary guy, there is a shift towards the slow light oscillation levels or towards the red side of the visible spectrum. It's like the photon stream is stretched to keep both viewpoints seeing light properly in "their" reference point. I know this seems bizarre, but this has been proven so please understand that your consciousness affects light.

Let's Go Faster

Just for kicks, let's see what the light will look like if the difference in speed is really, really high. Let's say the guy shining the light in the last example is going close to the speed of light himself. The apparent wavelength would get longer and longer and appear to be radio waves so we couldn't really see a light shining at all.

The thing to understand from this section is that light is a personal thing in your personal universe, while photons are shared-- light is not.

If you don't learn anything but the above, you will be miles ahead of most people today.

What is Light REALLY!!

I need to broaden your vision of light. Not as broad as Walter Russel's, but more than you typically think about light. For that expansion, let me first restate the Biblical description.

1. There is a seemingly crazy group of verses that tell us that God created **LIGHT** on the earth during Time Period [or YOWM] number one--- well before he created the Sun to make the light. That sounds like the egg before the chicken thing, but there it is in the 1st chapter of our Bible. [I know you read that YOWM means "Day" and that is sort of true. We could say chapter one deals with the DAY of Light, but that CERTAINLY did not mean a 24-hour time period.]

2. The book of Genesis also tells about how God Created a great **LIGHT** called the Sun to signal the daytime 4 "time periods" [or YOWMs] after he started rebuilding the Earth as required because of a terrible war that the prophet Jeremiah tells us left all the cities destroyed and the earth without form and void.

3. The Bible also states that when the rebel watchers lost the Heaven War something very special was taken away from the rebels. They were turned into humans as one of the punishments, but the second one seemingly even more

severe. The Bile indicates that the **LIGHT** was taken from them.

4. Another place Jesus said he was the **LIGHT** the truth, the way.

5. Jesus also said, "Believe in the **light** while you have the **light**, so that you may become children of **light**."

These things sound like Moses and the other writers had no idea what they were writing in Genesis and other parts of our Bible. I believe that God was trying to tell us something about our universe and what LIGHT really was. After all; if you can't trust God, who can you trust? In the first example, the word light can certainly be interpreted as life [of some kind], the second example would be what we normally think of light made of electromagnetic waves; and the third example, we come back to a definition of light that has to do with life again.

Light and Life

Hopefully, you can see that, somehow, light and life are connected together just like potential energy and kinetic energy are linked. It would have been nice if Niels Bohr had finished his work on how light and life are interrelated, but you'll simply have to make do with mine for now.

There is a symmetry of light to be considered. One might say that the outward expression of light causes the inbound condition we might call darkness or "inverse-light", this secondary light described in the Bible and other ancient texts ties the "inverse-Light" to human life.

You need time to think about this so put the book down and walk around for a moment.

Have you wondered about the similarity of magnetism and gravity? Well, I have and there is this same similarity between Light and Conscious Life as well. We need to recognize this link as we go further because light references are pretty hard to follow because many things get mottled in figurative speech.

I think it is safe to indicate that light has a duality with living. I think we can feel the similarity and certainly the idea of death and darkness have a similarity seems reasonable---just because.

 If Gravity [secondary dimension in the Particle dynamo] is associated with Magnetism [secondary dimension of the operational dynamo] then it should not be a stretch to say that light [a tertiary dimension of the operational dynamo] is associated with the Soul [the tertiary dimension in the ethereal dynamo]. As the physical word "light" increases, the intersection of the soul decreases. With the new revelations and something very strange we call light, let's check out the LIGHT ANOMALY.

 I hope you are seeing that perpendicular life and mass go along with the lateral time descriptions where light and life are exchanged perpendicularly.

Light Anomaly and Description

This whole perpendicular matter observation of light is pretty important when discussing the errors in Einstein's Energy formula written around light speed. This book is not a mathematical listing of numbers that add up to some esoteric characterization of matter that no one can dispute. It is also not a complex discussion that takes a scientist to understand what in the world I'm talking about. This book doesn't, necessarily, standalone. Instead it is a discussion of discussions and a theory of theories, so to speak. Hopefully, by looking a wide assortment of ideas and bits of evidence it clears up many of the anomalous details presented by other theories of matter.

For instance, the atom is not the essence of matter just like a photon is not the essence of light. They are simply EASY models that show some of the characteristics.

There certainly are atoms or at least atomic characteristics and characteristics of photons, but basing matter on atoms and molecules and nuclear forces and covalent bonds and all the other things presented to you previously will not allow you to understand more basic elements of life. All matter should be thought of as being made up of vibrating nothingness. This fundamental characteristic will change how you look the entire world. I know you think you are comfortable with the way things are, but let me push you a little harder and reinforce some of the elements briefly described so far.

Vibrations of Light and Life

We know that the faster the photon of light thing vibrates the more powerful it becomes. Soon the fast vibrating photon thing becomes dangerous to humans as it can go right through the body [x-ray] and if it slows down too much it changes into something we call radio waves. I know you are thinking that these radio waves must not exist because they don't produce light and they have no mass, but let me assure you that sometimes these photon things do act like normal matter. If you look at the diagram following, there is a wiggly line. The faster wiggling represents a prime particle vibrating faster and faster. Radio waves turn into light that turns into the deadly gamma rays.

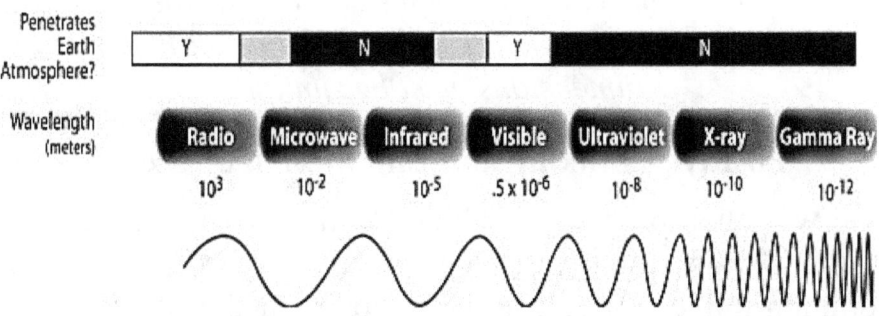

Today, the chart can continue even farther in both directions as light starts its vibrational journey as static electricity. As vibrations start to increase, we call the outcome electro-magnetism. When we get to the highest vibration, we can define the vibration or thing as pure Kinetic Magnetism. All three dynamos have similar static and kinetic limits and dualities with the other 2-dimensional dynamos.

As the ethereal dynamo goes through the same type of transformations from the static carnal life to the kinetic soul life and matter goes through this same transformation from the static Aether up through the pure kinetic gravity of what we call a black hole. We must get a sense that the 10-dimensional universe concept seems to pull everything together.

Resonance of light and Life

It is reasonable to assume that ultrahigh frequency light, black holes and the soul are joined by vibrational similarity as are the complete carnal life of a tree, the Aether that could eventually make matter and the potential electric fields that could eventually do work. I will get into this relationship more as we investigate life in more detail, but here is where we are right now,

For light to provide a resonance and be sustained, the surrounding vibrational characteristics of the other dimensions must be similar.

Positive Thinking and Elimination of Gloom

One way this might be attributed is that when we sense the light as "warm or comforting" it allows us to witness this common universe because the majority of the life-forms are resonating at a vibration level that allows these frequencies to affect us in this "Warm" way. If we vibrate more slowly, things will appear to be darker and more foreboding, even in the middle of the day. If we can vibrate faster, awareness of a higher level of light can bring comfort, insight and understanding. This sounds like all that "power of positive thinking" stuff and it does have some similarity, except that vibrational base is what we call life instead of light. [I told you, life and light had a similarity and this is one area that it can be noticed.

What I mean by this is that visible light is a moving thing. It goes beyond sight, and extends into comfort, desire, understanding, and awareness. Light even can explain matter because all dimensional dynamos are intricately locked together. As one dimension is stressed outside its normal resonance, the others MUST follow suit. With that, let's look again at Einstein's struggles.

Einstein

As we discussed previously, one of the first people to know there were problems associated with the "Clean" atomic theory was Einstein. He knew that if the electron cloud around a Protonic nucleus was to be held in place, there must be an unseen group of <u>things</u> between the nucleus and electron particles. John Keely had called this an "ethereal force" so Einstein decided to describe what he called Aether in keeping with John Keely's observations. Let's look again at Einstein's observations and look for commonness between the structural dynamo associated with Aether and the Operational dynamo causing light to be sensed.

Einstein's Squishy Things

Like many others, he called this collection of things the "Aether" and he said that it was made up of invisible, sort of squishy, particles that bounced around between other charged particle systems such as the electron and proton. These "Aether" particles are invisible and have no mass, or at least none that we can sense. That was only the beginning. More and more data were collected which allowed a better understanding of the dynamics of the atom and even the electron itself. Einstein's Aether was actually starting to make sense, but Einstein didn't stop there. He couldn't.

> *After all, he had now completely shattered his E=MC²*
> *formula with these massless particles. "Who Cares!", he*
> *might have said as he was on a roll!!*

If you remember the energy equations from before, all were of the form $E = 1/2\ AB^2$ so his universal law of mass-energy should be re-written.

- $E = \frac{1}{2}\ MC^2$ [possible universal law of mass energy that goes against Einstein's original one.]

Does it make sense that the mass energy model would be different that all the other energy models? No! Everything seemed to fit in the old $E=MC^2$ until Einstein's invisible squishy things came along. Let's assume that the added ½ in the equation must be accounted for. What we find in Einstein's squishy things concept is the rest of the elusive picture with all matter and apparent affect. To make things worse, Einstein's work indicated that mass going the speed of light would become infinite, but photons going the speed of light [and faster] don't have infinite mass. What is going on? The equations are, simply, augmentations of a unified particle.

Unified Particle Theory

Einstein didn't change his original mass energy equation, because it seemed to work for most things but he did conclude that there must be an error, which allowed for one unifying particle rather than the hundreds that were continuously being found inside the atom. This unified particle theory is the basis for the new atomic science that I have been preaching over the last few books and it helps us define light.

The NEW SCIENCE is looking into the details of the elemental parts of atoms. The current studies are getting close to finding that single particle that makes everything. When I say everything, I mean everything including matter, light, and gravity. Because light is so associated with mass, possibly a better understanding of mass will allow for a breakthrough. Let's look at some of the tiniest known particles. I apologize for the details, but it will come together in just a little bit. Some of the details were covered to one level or another in the first book of this series, but I think, we should review them as we expand the concepts to this very strange component of life we call light. [I hope this section will start bringing things together for you!]

The Huge Electron

An electron has a mass and movement; therefore, it has a gravity and electro-negativity. Scientists call the combination of these features "Electro-magnetivity". The electron can be further diminished into particles called "quarks" and "positrons", but as we go farther down into the individual elemental parts of the electron, the basic component, which we now called the "Boson" contains multiple fermions [quasi-masses missing components that allow them to be visible.] and **a frequency**. The reason for saying quasi-mass is that fermions making up bosons are invisible and Bosons made up of invisible fermions **can become invisible** just like a photon is before it is excited, just like a graviton is when it is sensed, and just like neutrinos when they are generated. By placing a number of these "bosons" together, the apparent mass becomes a "quark". This quark thing then produces either, or both of the mass forces of the universe, gravity and/or electro-negativity. Put enough quarks together and

82

you will get the huge electron which can change electro-negativity into electro-magnetivity which is a whole different type of force and is part of the definition of light. Let me review this string of events to the tune of that bone song.

Aether connected to the Fermion as vibration begins
Fermion connected to the BOSON [A Particle is born.]
Boson connected to the Quark thing [Electro-negativity is born.]
Quark thing connected to the Electron [Electro-magnetivity is born.]
And the atom's not far behind.

Light is somewhere in the mix, but it is more special that the standard run of the mill fermion. Another way to look at light comes from the Tamashii model of the atom that I described in the first book of the series.

Tamashii Light

Everyone knows that if you put two frequencies together, the combination is more than the two frequencies. It also contains "beat" frequencies that are associated but different that original two. According to the Tamashii Model, collisions of these two "different frequency" particles do the exact same thing. Interactions can cause secondary "Beat" frequencies to be generated. This "beat" frequency is the photon emission [light]. So, in general, the close proximity of two different frequency particles is what causes light. Later on- Milo Wolff described the vibrational collisions as exiting vibrations and vibrations entering the universe from a "Linked" one, but the Tamashii model is good for what it was.

Just think of light as an illusion of vibration. Remember that you can't really see light. All that the brain gets fed is the vibrational content of these photons. The brain changes the frequency of the

83

This is a little easier to believe than- "sometimes light is a massless electromagnetic wave and sometimes it isn't because it has mass and is a tiny particle." While this is the "Normal" way to describe light, it really is a stupid way. That brings us to quarks.

Quarks

Changing the frequency can change the characteristics and union possibilities. What this means is that the huge particles we previously thought of as building blocks of matter, atoms, are made up of still smaller particles that can change because of some outside influences. This "change" can and often does include invisibility or the output of this light stuff.

As we get inside the atom, if 3 or more of these tiny, tiny particles, called Bosons, join, they become one of 6 different types of Quarks [Charmed, Strange, Up, Down, Beauty, or Truth]. These are weird names for weird particles and I didn't not name them. Certainly, I wish there was an actual TRUTH quark that would tell us what matter, life, and light were, but we'll just have to keep investigating. Odd quantities of Boson particles are "apparently" visible and appear to have mass, even numbers have no apparent mass because they are electrically neutral and don't tend to interact outside themselves.

Gravitons

One such particle is the graviton, which has no apparent mass, but has a gravity, which requires mass. [Ahah!! We are talking about things that are so small that we can barely

84

imagine them and now we are trying to define the universe with them.] This "theory" ties together photons, wave theory, particle theory, gravitation without mass, and these new discoveries of molecules changing characteristics. We at least have a chance at having a real definition of particles with the Tamashii Model rather than the usual "definitions without details" we tend to accept. When I say "usual", I mean when a text book simply states "therefore it can be concluded" and you read before and after the statement and there is no way that the result can be concluded except by defining away many exclusions and exceptions.

A photon is NOT sometimes a particle and sometimes a wave like you were told. The quantity of particles collected by the photon and their frequency characteristics make them invisible or visible depending on what they come in contact with. Mass, really does not exist in the true sense of the word.

Let's say there are 2 fermion [quasi-mass] things minding their own business and they happen to be vibrating at different frequencies. What happens is that there is a beat frequency produced and the particle become visible in a string with that vibrational pattern, but there is something left over. If 2 frequencies beat, there is a negative and a positive beat frequency [known as a positive and negative beat frequency one becomes visible as a particle while the other generates electro-magnetism or a PHOTON. The diagram following shows how this function might be interpreted in a "time-space" way.

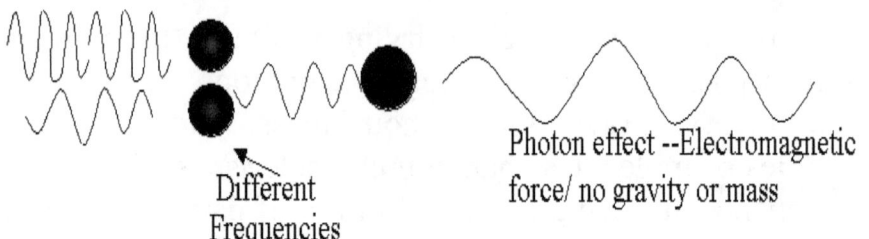

Different
Frequencies

Photon effect --Electromagnetic
force/ no gravity or mass

These out of phase vibrations can be considered time inverted vibrations or vibrations associated with in-waves, or even waves from outside our universe. One could define a graviton as a sub-particle that is enveloped with an in-wave from another universe, but some people don't want to even know about in-waves. Therefore, let's see if these sub-particle things can be defined without an adjacent universe.

Particles

While it attempts to define particles, the Tamashii model is very limited and it does not define light by itself. What it does is open up our minds to possibilities necessary to continue so that some of the other weirdness I have been describing doesn't sound as strange. I wish the existence of light was even as simple as the previous discussions of all the fermions building quarks building electrons and so on, but the deeper you go the more confusing it becomes. Not to be thwarted, Keely, Bohr, Einstein, and others went deeper into a relationship between things that exist and things that don't quite exist, including this "light" stuff. This "Ethereal force" Keely came up with involved things we simply could not see at all.

Things We Cannot See

After I initially spewed all of this out, I'm sure you got confused so I brought in the other 3-dimensional elements so that you could start to see something. By the way, there are many normal things we cannot see.

- **Gravitons** are particles that have a mass, but not gravity so we can't see them directly.

- **Photons** are sometimes particles and sometimes the particle component disappears. That doesn't mean light doesn't exist. It simply means that sometimes mass must be associated with other things to allow us to "Understand" them. What this really means is that light that we cannot see is responsible for seeing things. How odd is that??

- **The Event Horizon of the black hole** is where mass is miraculously generated even though we have a law that says, you cannot create matter. The matter must have already been there.

- **The whole black hole** can't be seen because it has so much gravity, light can't escape. If we remember the Laser example, light is emitted when an electron is pushed away from its associated nucleus and when the electron goes back to its normal the photon is generated. We can assume that the electrons NEVER are allowed to move away from their nucleus's in a black hole. In fact, the electrons are forced closer to the nucleus all the time. One could characterize a black hole as feeding on photons, but really they are simply "time challenged". They are no longer associated with our reference time, so to speak. We are also told that there is a possibility that a black hole can hold the key to travel to another universe.

We can use this additional data along with other details about the universe to helps us define this elusive Light. Especially if we start with the generalized definition of a universe I have been building during this series of books.

A universe is actually a "perceived" system of systems that enjoy a unifying stability. Each of these systems is a dynamo characterized by mutually perpendicular dimensions.

The Electro-magnetic, Fermio-gravitational, and Self-spirit "systems" are indelibly linked. Therefore, light which is associated with the electro-magnetic system has a dual in the fermio-gravitational realm and also in the self-spirit arena.

Before we get into the life part, we must try our best to establish the link fermionically. One could say that light is identified in the particle world fermionically. We can describe a fermion as mostly a static structural characteristic, but having a higher vibrational component than the Aether. We call the fermion equivalent in the operational dynamo the photon, but I hope you know by now that light is not simply photonic. Let's investigate the world of the quasi-particle and see that defining matter isn't that simple either. I'm sorry for repeating and saying things that seem so odd to you, but you are going to have to relearn everything you took for granted if you have any chance at understanding something as difficult as light.

Fermionic Particles

The following list is a portion of the "not quite particle" set defined mostly from the structural or the Operational dynamos. As I have stated many times now, a fermion is a particle that has no apparent mass and exerts gravity and/or electro-magnetism or something else that allows us to sense them. They are not mass, but at the same time, they are mass. Some have tried to define entire particles like Electrons and Protons as fermions, however, most, today use the fermion name as a particle missing a component needed for it to really exist like a "graviton". It has gravity, and must have mass, but no one can find its mass.

Neutrino	This is a quasi- Lepton that is a component of an Up-quark, three types known [Electron-neutrino, Muon-neutrino, and Tau-neutrino.] **They have almost no reaction with matter and can pass through the Earth**--- They have no apparent mass.
Electron hole	A lack of electron in a valence band. While everyone uses these electron holes, to interpret electricity, they don't completely exist.
Photon	This previously was considered a Boson that has no apparent mass but has electromagnetic properties. It can be "modeled" with 2 quarks or equivalent particles. It exhibits no Gravitational force, but instead it SOMEHOW, makes light. Like many of these vibrational components, the vibrational travel is faster than the speed of light.
Graviton	Like the photon, this is another fermion possibility that has no apparent mass, yet it exhibits a strong gravitational force. This suggests that an even quantity of quarks is combined in its makeup.
Chargon	A quasi-particle produced from an electron spin-charge separation
Configuron	An excitation in an amorphous material associated with breaking of a chemical bond
Exciton	A bound state of an electron and a hole

Fracton	A collective quantized vibration on a substrate with a fractal structure.
Holon	A quasi-particle resulting as a result of electron spin-charge separation
Libron	A quasi-particle associated with the motion of molecules in a crystal.
Magnon	A coherent excitation of electron spins in a material
Majorana fermion	A quasi-particle equal to its own antiparticle in superconductors
Phason	Vibrational modes in a quasicrystal from atomic rearrangements
Phonon	Vibrational modes in a crystal lattice associated with atomic shifts
Plasmon	A coherent excitation of a plasma
Polaron	A charged quasi-particle that is surrounded by ions in a material
Polariton	A mixture of photon with other quasi-particles
Roton	Elementary excitation in superfluid Helium-4
Soliton	A self-reinforcing solitary excitation wave
Spinon	A quasi-particle produced as a result of electron spin-charge separation
Gluon	A quasi-particle that can cause an interaction between Mesons and quarks opposite to gravity.
Lepton	A quasi-particle [no-mass] that does not attract nuclear force.

OK! We have a lot of very loosely associated quasi-particles running around our universe that do not help us interpret light. However, it is important to understand here that there is a separation between light [the electro-magnetic characterization of visibility] and the photon [the particle understanding of visibility]. We will also find that there is a "light" description as well.

I told you that light was not simple, didn't I? Maybe now you will believe me when I say something.

The main reason I put this list in the book is to show you that what you and everyone else knows about existence is pretty much a guess at this point. Trying to determine what things are by examining the "Particles" cannot be a conclusive study because so very much of the "Essence" of matter---- ain't matter at all". Light, not only is not matter, there is a substantial problem in our universe.

If Light continues outward from its initiation point to the limits of our universe, soon the light would simply be too far from the "normal area" to be used and the universe would begin to dim until nothing we would call light could be sensed.

Photons don't seem to go away. Possibly we can classify life in terms of one of these quasi-particle photon things. I know that sounds bizarre, but what if we shift the direction of time? Photons [or more precisely electro-magnetics] possibly should be characterized in that lateral time I mentioned before. I think you will see what a mean in this next section as we dive deeper into lateral time.

Before we get to lateral time again, let me state the previous description as it relates to life. It may sound similar.

If Life continues to trek forward in time from its initiation point to the limits of time, soon the essence of "Life" would simply fade away. No more life could be generated. Entropy would be satisfied and the memory of living would be forgotten in time.

Like light, life doesn't seem to be going away.

More Lateral Time

Don't worry about time, light or life going into infinity never to return here. Certainly, that is something to worry about and generally the way many think of time just like they think of light simply going on indefinitely to infinity ever to be lost in our universe. Certainly, Einstein and others determined that mass had no end and went on to infinity ever to be lost and it worried Einstein. All this expansion to the limitless darkness has plagued physicists for years. Well, forget about that right now and let's shift time itself again. As we do it, we review this concept we will get a better image of what light might be.

For this discussion I will have to pull down this whole concept of time that you currently have some more. Basically, we can either look at time as a constant, as this impenetrable limit, or as something else. In this chapter I am going to bring out more of the "something" else. We can easily imagine time going forward with the past being behind us and, potentially, we can understand the concept of backward time that I discussed in the last book, but what I need to expand here is sideways or lateral time. This is not to mess with you, but because several areas in the discussion of light requires a limited understanding of this lateral time.

Vibrating Time

Before you can really understand time, you must first see it as one of the dimensions that makes up what we call a universe. As such, it must be made up of the thing that makes up all things.

NO!

I don't mean atoms or bosons or even the almost unperceivable fermions. Time is not made of particles. Hopefully you already guessed it. What I really mean is *vibration*. We accept the vibration of light and the other things around us, but we typically view time as this straight-line thing starting at the time we are born and ending when we die. We can sort of extend this same "time-line" from the beginning of the BIG BANG thing until the end of all time, but it still is in the same direction and has the same constant/linear dimension. Oh!! How comfortable this description is; very visible and very easily described.

Vibrating Electricity and Light

If electromagnetic fields didn't vibrate, they simply would not exist as would light itself. What I mean is that if you ever stopped a photon, you would be holding nothing. The bad part of this stopping vibration is that if we stopped vibrating, we would no longer exist as well. If we look at atoms, current studies indicate that they are simply clumps of common vibrational nodes rather than true substance. Even though everything can really be explained better vibrationally, let me first straighten out the time line so it looks more like what we perceive.

As I previously touched on, if we view time from the side, we can see the beginning of time and the end of time along a line in front of our eyes. I've labeled the viewer as God, because he may be the only one that can perceive this thing. From this vantage point, everything that happens to you from the birth to death are all shown up in one instant. There is no future or past, there simply is.

Normal viewing of time

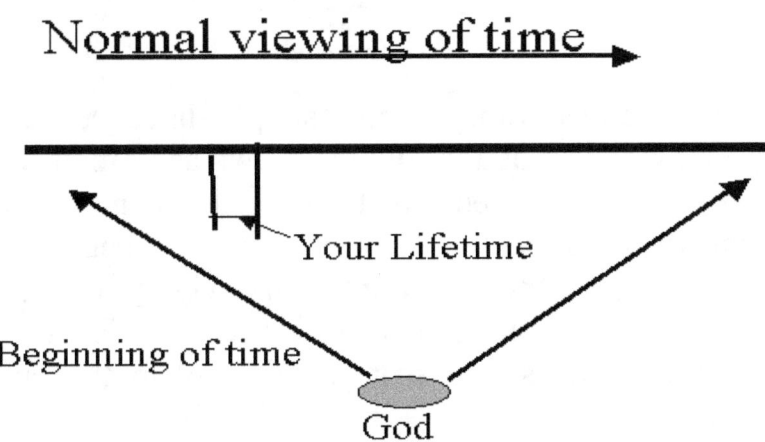

Your Lifetime

Beginning of time

God

As I mentioned at the first of this section, time may be vibrational like all the rest of the dimensional strings of this universe. Instead of a straight line, on the following page, I am showing it vibrating just like everything else does. As we go through time, the hills and valleys don't mean anything to us, but the variations could be witnessed laterally. The hills and valleys might be certain cyclic pressures like destruction periods, Ice Ages, wars, and other things that mark the cyclic nature of time and God could look at all these peaks simultaneously.

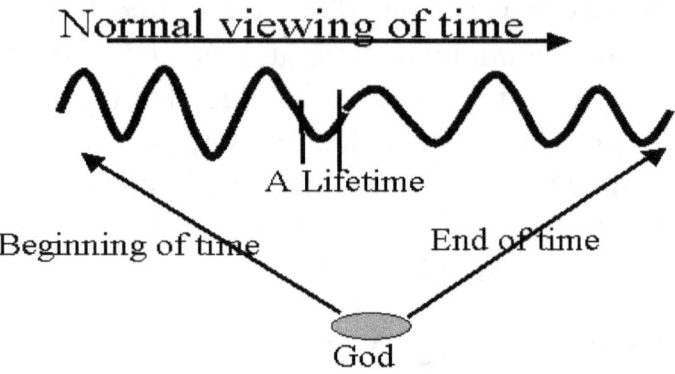

I haven't changed time here; I simply have changed the viewpoint. Notice that a lifetime is shown as a small segment of what would be viewed. The beginning and end of time are only shown for direction. There may be NO beginning and NO end for all we know, so think of the wiggly line going on and on as far as you can see or in a circle as depicted in some examples of dimensional strings that explain the quantum effects noted in life.

Light Seen Laterally

While lifetimes take up a section of the time line, typically, light goes the speed of light. It is here one instant and gone the next only to be regenerated and be found again for another instant. Light doesn't actually travel along the time-line. As Einstein predicted at the speed of light there is no time. Light would view the universe LATERALLY as shown in the next graphic. Light would appear the same as life appears when viewing time normally. Normal mass would have no apparent mass equivalent and light would be represented as having this equivalent of life.

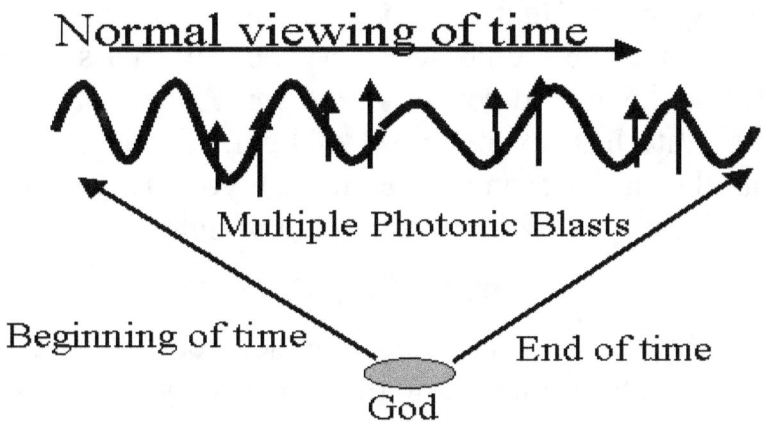

Normal viewing of time

Multiple Photonic Blasts

Beginning of time

End of time

God

Speed of Light Example

That was the easy description with no meat. Let's put on some meat and see what happens. If a person leaves here in a rocket going the speed of light and returns going the speed of light, what would the rocket look like? The answer is that the rocket would gain infinite mass along the direction of travel. It would look like a beam of light and it would not travel on the "normal timeline. The rocket and the person experience LATERAL TIME as shown below.

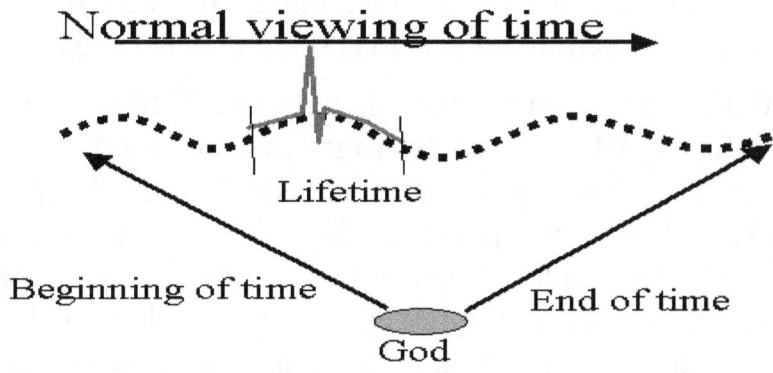

Normal viewing of time

Lifetime

Beginning of time

End of time

God

The time-line was expanded a bit to show detail the spike in the middle represents what happens when the rocket goes close to the speed of light. If a person could see what was happening outside, he would see everyone's life passing in an instant. The downward portion of the spike is his return to normal home at close to the speed of light. If you haven't seen it from the earlier examples, let me tell you that you will turn into--- "light" ---- if you go the speed of light.

Turning Into Light

Let's look deeper. If you could see someone who was viewing you in lateral time, how old would they get as you aged? Of course, they would not age a day because they could see your entire life as an instant. If you go the speed of light, how old do you get with respect to those not going with you? The answer is that you would not age.

A more defining answer is that if you could possibly go the speed of light you become light and are traveling in lateral time. However, all the particles in your body are vibrating so the particles making up your body are going backwards in time or they must stop vibrating.

Forget I said the backward time thing and let's move on. That was discussed in the previous book of this series.

If particles stop vibrating, they do not exist. While this would be bad for 'LIVING" people. Light may have no reason to exist or not exist. Therefore, crossing over to adjacent universes should be possible for light. The reason I highlighted the "Living" part is that our "spirit" [one of the Ethereal dimensional components] may be able to do this transfer without issue. It may reside in a place where lateral time is the forward time and forward time we perceive is

98

what would be considered lateral time. Let's look at the diagram that was presented before.

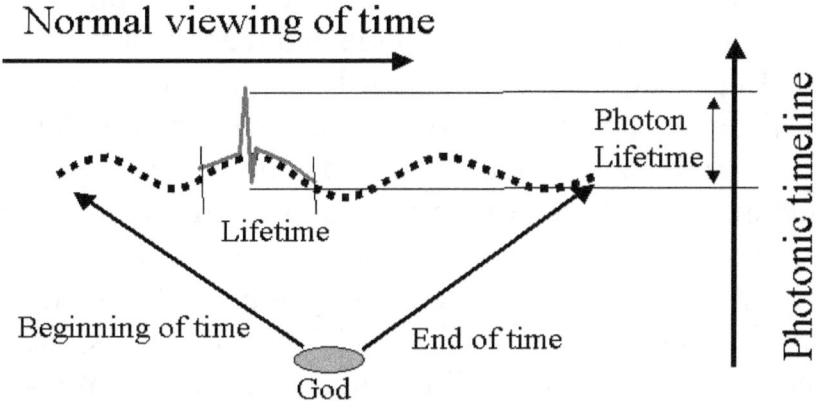

In the diagram I showed that light really had no normal time component but would look like a spire out from the normal timeline, in fact, if we were to recognize the lifetime of a photon, it more correctly is determined as shown above. Beginning and ending in the blink of an eye, but it could last for the equivalent of centuries in what I'm calling lateral time.

Speed of Light Example

In a lateral-time-world, what would a rocket look like? The answer would be that rocket would look as long as the entire trip the rocket took in time [sort of converted to length]. I brought that up for a reason. What would a rocket going the speed of light look like in "normal" time? You guessed it; it would be as long as the entire trip to a viewer that was stationary.

Vibration

If vibrations are emanations of modification over time, then the rocket could be viewed in lateral time as a solid mass. I know that sounds like a black hole and matter must be being sucked into oblivion, but it's not. Remember we are looking at a different dimension. To the viewer of lateral Time, all the vibrational motion would now be produced simultaneously. As a particle moved over a distance, it would look like a line drawn across many more lines. Let's explore this just a minute. How do we perceive the vibrational patterns associated with a mass? The answer is that we perceive them as a mass. We cannot see the vibrations and, frankly, we cannot even understand them as vibrations. It is as if our "impressions" of things are associated with viewing matter in lateral time rather than linear time that we experience.

Photonic light would be seen as solid beams like wire and here is another thing to think about. How long would the light beam be visible? Da-^{dah}-dah-$_{da}$-da-^{dah}-dah [I'm humming the Jeopardy tune.]-----"What is forever?" is the correct question.-----I mean the image would be there forever in lateral time.

Lateral Time is the Speed of Light

I know we use this "speed of light" thing every day, but now we can define it differently and possibly more accurately.

The speed of light is when vibrations become solid. It is lateral time.

I know you were expecting that absolute zero or minus 473 degrees would make all vibrations stop, but we actually had to go in the other direction.

The speed of light also is the transition from linear time to lateral time.

Once we cross that boundary, time travel, by this description should be possible in forward and reverse by simply injecting yourself into the lateral picture presented in lateral time.

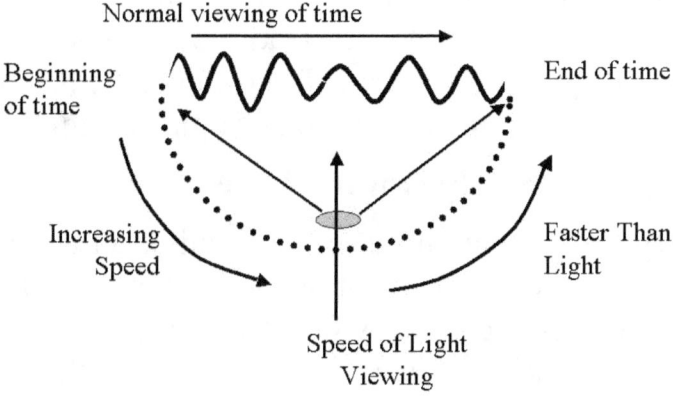

The pervious drawing, hopefully, will sort of explain what I am saying here. The curved arrow on the left shows what happens as you move closer and closer to the speed of light. What you would see is that everything would speed up for you. As you approached the speed of light things would be skittering around faster and faster. Soon there would be a blur and everything would be impossible to understand because you would be seeing all time in an instant.

Beyond the Speed of Light

The blur begins to slow down slowly as you go faster than the speed of light as described by the curved arrow to the right of the drawing. What you notice is that things are going backwards now. Soon you can get to a level that is similar to where you started with respect to the motions of everyone, except that everything is backwards. While in this state, one could gain knowledge of the future that would now be represented as the past. Once the information is retrieved, slowing down to a stop brings you back again to "normal Time" and you have the secret of the future with you. If you noticed, as soon as you go faster than what we call the speed of light, you don't go faster. Instead, you go backwards and that brings us to an important description of time which is certainly needed for any discussion of light and life.

Backward Time

Let me tell you something else about this backward in time vision. You would be seeing an adjacent universe rather than our own. I'm not even sure you would recognize any of the elements of our universe very easily at all, but that is one of the subjects of the previous book in this series. For us we only need to understand that there is something we can call a Constance of Time.

Constance of Time

Just like all things, light, matter, and life, time CANNOT simply go into infinity and be lost. If time was only one way, soon we would have no time. Therefore, there is an underlying negative to time. We can perceive this negative time as an adjacent universe [some call heaven] where time is completely backwards. If we view this NEW world, their in-waves are our out-waves, their electro-magnetics is our fermio-gravitations, etc. There is a strange linkage between our universes and both are needed for either to survive.

I'm gonna let you think about that a couple of pages because I'm pretty sure you are wondering what's going on and your brain needs some time to put everything into perspective. When I get back into it, I will ignore this secondary time dimension and get back to the more direct discussion of resonance.

Blank Space for Resting

Another Blank Space for Resting

Use the time wisely and do not skip this page.

Another Look At Resonance

If your brain is through resting, let's get back to it. In electro-magnetics, resonance has been used and studied for years. Simply stated, the resonance thing is the most comfortable frequency for an electric and magnetic field to stay at. It is a point where electric and magnetic fields both have the same strength and when that occurs, the effect of the 2 fields is most noticed. In one regard, we can look at resonance as the tertiary dimension of the specific dimensional dynamo that it works from. Well, the same thing happens in any vibrational structure, this same feature applies. Let's look at the fermionic and gravitational resonance.

Fermio-Gravitational Resonance

I'm going to get into what makes the various atoms different so listen closely.

The difference between a helium atom and a gold atom is vibration, but what keeps the gold atom together? The answer is resonance.

Just like the electro-magnetic resonance, particles express this same feature. Particle resonance is the most comfortable frequency for a fermionic and gravitational field to stay. It is a point where fermions and gravitational fields both have the same strength and when that occurs, the affect of the 2 fields is most stable. If these two fields are stable at a high frequency, they appear to be a large atomic mass. At lower frequency resonances, the lighter atoms become apparent.

Resonance Example

Crystals provide the best example of structural resonance. A crystal likes to vibrate in accordance with the lattice structure of the crystal itself and its thickness. By placing a small amount of potential across the crystal [by either mechanical stress or electrical stress] one can make the crystal vibrate at its preferred resonant frequency. To make the vibrational frequency higher, simply make the crystal thinner. Thinner and thinner crystals go higher and higher in frequency. It would seem that if one could get the crystal to be only one crystal thickness that the thing would want to vibrate at close to the frequency of solid gravity. Of course, this isn't the case as there is a limit to how thin you could make the crystal. A black hole would require the crystal to be thinner than a single atom and the crystal would fall apart, but certainly a super thin crystal would have a super high frequency resonance. From this example one might think that the thinner an electromagnetic field became, the higher would be its characteristic resonance. Life also should be investigated.

Life-Consciousness Resonance

You might have gathered from the similarities of the various examples that the esoteric components of life and consciousness would also have this resonance feature and its manipulation can be described and so it does. In this case, a

107

life force and consciousness level would be matched to provide the most stable life pattern. Those attaining the highest level of consciousness would naturally produce the highest level of living, but this gets into something Indian Gurus call chakras. I talked about these chakra things in book 2 of the series, but the thing that is more important in this book is how resonance affects life and how we can affect resonance.

Resonance Consideration

We may consider the resonances of the three-dimensional dynamos as an indwelling percentage of existence between this universe and the adjacent one. In book one of the series this idea was introduced as the theory of super-symmetry. All things in this universe are indelibly linked to our associated universe such that as things get larger here, they get smaller over there. This isn't my theory, but I'm taking it to a different level here. Not only does that mean that if we create a mass here, we destroy a mass over there, it also means that EVEYTHING is linked to offset the characteristics of the Law of Entropy.

If an electro-magnetic vibration resonates at 1/3 the speed of light here, there is an equal and opposite vibrational component that is associated with 2/3 the speed of whatever light means over there. If Milo Wolff's in-waves and out-waves are truly symmetric, then we might assume that matter and electromagnetism are reversed in this unusual world. While many other abstracts can be ascertained, one which helps me understand how Life and light are linked in this world is this entropy thing. For that discussion, let's look again as the structural dynamo.

Structural Dimension Entropy

In vibrational existence theories everything vibrates. The essence of matter is vibration, but there is an oddness to observe. The vibrations associated with the dimensions of the Structural Dynamo don't go anywhere. To get them to move through space one must add in the operational dimensions. We can think of it as getting bigger and smaller in a rhythm as shown in the graphic below. In the case of light, the bigger might be in this universe and the smaller might be somewhere else.

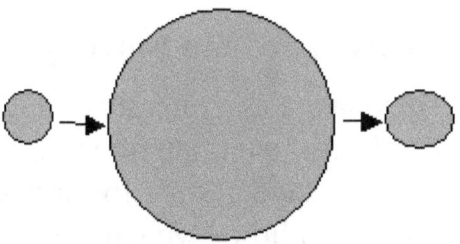

We could say the dimensions that build particles are not spatially motivated.

Fermionic energy is pulsing at the same time that its associated gravitational energy is pulsing. The nuclear energy element is pulsing right alongside, but nothing happens until other pulsing "things" come in contact with them and they either collect because there is a likeness in vibration or they are invisible to each other because they have a difference in vibration. In our universe, the operational dimensions push these pulsating globs to put them in contact with others to form matter. As the things are pushed together, others are separated identically in time, but not necessarily identical in space. If you remember Niels Bohr's quantum mechanics descriptions, you might see that everything in the universe is working in unison.

Law of Entropy

Fermionic and gravitational dimensions act inversely with respect to how the particle world uses these characteristics just like magnetism and electricity in the "electromagnetic or motion" world. This is where structural resonance or fermionic resonance comes in. As we look at resonance, remember we are really looking at an associative property with an adjacent world.

Resonance and Entropy

As I mentioned, resonance is typically thought of as sort of the sweet spot of an electromagnetic component. In electronics we use this sweet spot to find the channel we want to listen to on the radio or we can keep time by finding the sweet spot of a crystal used in an electronic clock. It is the frequency that allows the electrical and magnetic components of a system to have the lowest effect on the environment. That is all known and used by designers every day, but the structural dimensions do exactly the same thing as the fermionic and gravitational frequencies match such that they affect the environment the least. Having the least effect on the environment is the lowest state of matter or the most disorganization of matter.

We call the structural resonance the Law of Entropy. Everything tries to go to its most disorganized state.

While it makes no sense that an entire universe would be trying to disassemble itself, many experiments show that given enough time and little "outside" interference, things simply go away.

Law of the Nuclear Dimension

Einstein came along developed his now famous equation $E=-MC^2$. While Einstein didn't introduce the equation this way, this equation really identified the energy associated with nuclear force that counteracts entropy. What it actually is saying is that the force presented by the nuclear dimension [also known as nuclear attraction] challenges the other 2 dimensions that are being pushed by entropy. The larger the grouping of fermionic energy, the more the nuclear force contains it so there is no worry that a carbon atom will slowly disassociate into helium and hydrogen atomic clouds.

What About the Minus?

Oh! You saw the minus sign, didn't you? The minus sign helps us understand what light is. The minus sign simply means that nuclear force absorbs that energy or exchanges that energy rather than initiating "light". Here is where the fun comes in. The control dimension in the operational dynamo is light [going the speed of light] and the acceptance associated with nuclear energy is associated with this same speed of light characteristic. As atomic masses get larger, the

111

amount of "speed of light" components that can be transferred increases.

Hooray for the Minus

OK! That isn't helping. I Know. What will help is that carbon stays carbon against the law of entropy because something in our adjacent universe forces out enough out-waves [which become in-waves to us] to intersect the out-waves of the vibrational node identified with that carbon. So long as the adjacent universe emits, carbon will sustain. And entropy will be absolved.

Don't worry about waking up one day and being an infinitely spread bunch of quasi-particles so long as heaven is nearby.

Operational Dimension Symmetry

Many people work every day by manipulating the operational components of matter. They use electricity or magnetism and they listen to the electromagnetic waves associated with their favorite radio or TV channel. In this dynamo, I don't think I need to spend a lot of time on electricity and magnetism except to say that they parallel the structural components [i.e. Magnetism and gravity work in a similar way]. Electromagnetic resonance is the operational equivalent to entropy in the structural world, and like the nuclear attraction, light [or electromagnetic waves] act as the stabilizing component.

No Operational Entropy

In-waves from our neighbor continuously barrel into our universe. Operational Entropy doesn't really exist here, but our modifications of electro-magnetics here affect entropy in our adjacent universe. Luckily, almost all our manipulations of light is minuscule. We very rarely even break atoms apart to generate the larger amounts of electromagnetic fields. We very rarely pull apart black holes which would generate larger amounts of electromagnetic disturbances. We almost never even travel the speed of light to allow transfer from structural to an operational state. Our universe, currently, is a very orderly place. The largest disruption we typically are

113

associated with comes from the last of the dimensional dynamos.

Ethereal/Life Dimensions

While this is the most obscure grouping of dimensions, I don't think it will be necessary to bring in these dimensions to define light in a reasonable way. We cannot define light without this dimensional dynamo, but, I need to hold it until we study life and I don't want everyone to close the book wondering about spirits and souls and the like. If they close the book at the end of the book, that is fine. Keely's Ethereal force was noted specifically as many of his experiments would fall apart if others attempted the same. While he remained frustrated with this unknown Ethereal thing, we will try to put definition around it so that we can understand living in a broader sense.

Ancient Light

As I mentioned before God made light after he had created mass as was identified in the 1st chapter of Genesis and finally he made life. Then I said he put some of this light BACK into the descendants of Adam. Except for a little tweaking, the entire world was going after that. Possibly, we can gain some understanding from reading more than just Genesis. Below are a few insightful texts in no particular order that may open our eyes a little. I know I brought them up before, but bear with me.

Biblical Light

"Genesis"- *"God said let there be light and there was light"* **[This was before the sun light was made.]** In the catholic "Prayer for Enlightenment" we find *"O Holy Ghost, divine Spirit of light"* **[The holy ghost was introduced to allow understanding and return the light to humans.]** "Ezekiel" 1:4-14 tells us about flying saucers filled and powered by light.—" *Out of the midst of fire, of the whirlwind, came four living creatures with four wings. They had straight feet and they sparkled. Their wings were joined together. They did not turn. They all went straight and their wings were stretched upward. Two wings from each were joined together and two covered their bodies. Their appearance was*

116

like burning coals of fire. They went up and down and out of the fire came lightning. They were as fast as lightning.

Navaho, Pintes, and Hopi

There traditions tell us tell us about very ancient weapons made of light –*They all told of the Golden Strangers from the sky that came in flying canoes which were armed with Burning Rays of light.* [Possibly, Laser weapons were used in ancient times so there would have been a good understanding of light in general.]

Iranian Text

This comes from the sacred Zadspram- *"From the seed which was the ox's, they would carry off from it and the **brilliance of light** was entrusted to the angel of the moon in a place that seed was thoroughly purified by **the light** and was restored in its many qualities."* [This segment was after the watchers had corrupted almost all the animals. In order to reconstruct the animals, God had to put in more "Light".]

"Revelation of Moses"

This comes from the Gnostic book **"Revelation of Moses"** Chapter 33: 2-*"And she [Eve] gazed steadfastly into heaven and beheld a **chariot of light** borne by 4 bright eagles- and watchers going before the chariot. In this section* [Eve watched Adam being taken into heaven on a chariot of light after his death].

"Book of Abraham"

This second verse comes from the **"Book of Abraham"** chapter 4:3-*"And they said, let there be light, and there was light. The gods **comprehended** the light.* [This is an expansion of the Genesis statements. Remember this is

117

before the sun was remade. The odd part is the comprehend word. It seems that it is suggesting light was more than something visible that all could comprehend.]

"Origins of the World"

"Origins of the World" had this to say. - *"The troublemaker that was below them all destroyed the heaven and his Earth. And the six heavens shook violently; for the forces of chaos knew who it was that had destroyed the heaven that was below them. And when Pistis* **[one of the trinity]** *knew about the breakage resulting from the disturbance, she sent forth her breath and bound him and cast him down into Tartaros* **[Hell]** *and when they had become disturbed, they made a great war in the seven heavens. Then when Pistis Sophia had seen the war, she dispatched seven archangels to Sabaoth from* **her light**. *They snatched him up to the seventh heaven.* [This has the Heaven War, the troublemaker {Satan} and this odd thing called the light.]

--- *First she [Eve] was pregnant with Abel, by the first ruler [Adam]. And it was by the watchers that she bore the other offspring. -the first mother might bear within her every seed, being mixed and being fitted so that the modeled forms might become* **enclosures of the light**," [This seems to indicate that the losers of the Heaven War [sometimes referred to as watchers just like the non-losing angels and Adamic humans had children together so that the mixed breed might have the "Light".]

-------*"And he said, 'Come, let us create a man according to the image of God and according to our likeness, that his image may become* **a light** *for us.'* [The losers of the Heaven

118

War believed that man would somehow get them back the "light" that they lost in the war.]

His intelligence was greater than that of those who had made him. And they recognized that he was **filled with light** [**This was talking about some type of luminous essence in the descendants of Adam.**]

The blessed One, sent, a helper to Adam, **luminous** *Epinoia [holy spirit] who is called Life. And she assists by teaching him about the way of ascent. ----But the* **Epinoia** [**Holy Spirit**] *of the light which was in him, she is the one who was to awaken his thinking.* [**The spirit inside the descendants of Adam was somehow associated with LIGHT.**]

Cayce Confirms Genesis

Edger Cayce was a 20th century seer. The interesting thing is that many of his seeings have come true and some were extremely detailed so let's see what he had to say. -*"God moved and said, "Let there be light", and there was light. Not the light of the sun, but rather the light which—through which—in which—every soul had, and has and every had, its being."*

"Generations of Adam"

The Essene Jews were meticulous in copying ancient texts over and over. This is one of the most copied texts. "Generations of Adam" had information about LIGHT.[Chapter 8:13-14] *"A* **bright light** *shown from heaven illuminating the whole palace of king Canaan, and a mighty noise from heaven shook the air. Whence the palace stood there was only dust. A war broke out among the people. Armies devastated the land. The people suffered*

119

great desolation. There was destruction everywhere. [Horrible wars and weapons of mass destruction were known and the bright light from heaven that caused the air to shake really showed the power of light.]

In another section, we find more information. This comes from chapter1-:3-- *Leboa, Daughter of Tamar* [**Adam's granddaughter**], *devised a "Sword of Light" which penetrated the wall of defense around the city of Haner and began to drain the power from the wall"* [Here we find civilization well before the flood of Noah in a high tech war. Some type of Laser beamish thing apparently penetrated whatever the wall was and drained its "power". Sounds almost like the science fiction of today, but these weapons must have been amazing and some used light.]

Jubilees

The book of "Jubilees" is still considered canon in some orthodox Bibles of today. This comes from   chapter 2:9- *"Nor may we take revenge on him because he has **stripped us of the "light"**. He marked out the borders of the world and created man in his own image with whom he hopes **again** to people heaven, with pure souls."* [**Not only note that without the light, the losers of the Heaven War could not take vengeance on any of the heavenly host, they lost some substantial power without this light thing. Also note that the word "again" is put in the verse to let us know that man was here before Satan's Heaven War and it was recreated after it was over.**]

Greek Legend

In discussions about battles between the gods we find the following: *"Hot vapor lapped the titans, flames unspeakable rose bright to the upper air [outer space],* **lightning** *blinded their eyes."* [**Apparently light weapons were used in outer space.**]

"Vaimanika Sastra"

In India we find descriptions of light usually as some type of weapon. While "Viamanika Sastra" was last edited about 1600 years ago, it is strongly believed that it is an extremely old work. Much of the book details the art of flying and weapons used on these special transports called vimanas. Eight chapters are totally written to describe the inner workings of the flying machines from ancient times. Sometimes light was made into weapons the Aparoksha was a more benign use of light. *"One could employ this ability to* **project a beam of light** *in front of his craft to light his way."*

Brazilian Light

In Brazil, we also find a similar story from the Manacitas Tribe. They told and retold ancient stories about flying. One of their cherished legends talks about *the Macunbeiros, which were flying wizards that flew inside circular,* **luminous,** *machines.* [The indication of the craft being luminous may indicate an electric field corona just like those that have been observed in more recent sightings or it could have just been lights. The circular pattern with light all around is identical to the depictions of Ezekiel.]

Let's See What We Have

I know I haven't exactly defined light, but I think you know a lot more about it than you did. Here are some of the things I hope you have interpreted so far.

- Light was created before the sun

- Light was inside the watchers that fought in the Heaven Wars and it was removed as punishment when they lost.

- Before the worldwide flood, people were running around with laser weapons and blowing up towns, so there was a high level of knowledge about light in general.

- Light is somehow associated with an essence inside the descendants of Adam that is associated with transferring to the universe known as heaven.

- The survivors of the Heaven War tried their best to re-associate this Light thing by association with the descendants of Adam.

- Light vibrates faster than the speed of light, but its mass does not go to infinity.

- It is as if light travels back and forth between 2 linked universes.

- Huge amounts of energy are given up by light. By vibrating the light faster more and more energy is developed.

- Light has no gravity and no mass [most of the time]

- If light goes faster than the speed of light it must be going backwards in time.

- If light has a dual in an adjacent universe the dual would be darkness.

- Lateral time is the ability to view our world at the speed of light.

- When viewing mass in lateral time, time vibrates and other vibrations are halted.

- Light tracks a timeline that is adjacent to our normal timeline.

- If we view normal slow moving or stationary mass in lateral time, the mass goes to infinity. If we view mass going the speed of light in normal time, he mass goes to infinity

- Someone going the speed of light, never ages in normal time. He simply halts his travel in normal time.

- If a person goes the speed of light, and his mass becomes spread along the entire distance traveled, he becomes like light

- The unknown energy that is known as nuclear bonding seems to be similar to photon emission energy that forces the atom to contract.

- The force of the spirit seems to be associated with light. When people temporarily die they see a massive light, the holy spirit was identified as the light, and the heaven war rebels were punished by having their "light" removed so they would be stuck on earth.

- Light is the same as the dimension of Electro-magnetic force and it is perpendicularly associated with electricity and magnetism.

- When light mass disappears, it must reappear in an adjacent universe. This changing may be a doorway to an adjacent universe.

- In Lateral time. The beginning and ending of a lifetime can be viewed simultaneously, in fact, the beginning and ending of the universe can be witnessed.

- Light was the second of 3 creations of God, which includes something we call vibration, and something we call life. Each of the creations captured a 3-dimensional dynamo. Together the three make up our world.

- Biblically God created light well before he created the Sunshine and moonshine so original light was self-generated or nuclear.

- While the structural [matter] and operational [Electromagnetism] dynamos are substantially linked, the ethereal [life] dynamo is substantially separated with something termed as free will.

- In a universe adjacent to ours connected along the timeline would experience everything going backwards in time.

124

- Time dilation slows down vibrations and reduces the energy in a dimension. If it is not linked with others, the resonance of the universe [entropy] increases. Energy from an outside force must replace this energy or the universe will slowly run down. What we see today is that the universe seems to be more vibrant each year. Unfortunately, we cannot judge this as we are linked with any time dilation that occurs.

- Photons in this universe are photinos in an adjacent one. Photinos are photons going backward in time. There is always a symmetry between photons and photinos.

Let's Try a Definition

While there are a number of things we call light. The essence of what we use to build our visible universe is not and cannot be characterized without the conscious mind. Vibrational electromagnetic waves do enter our eyes and we can use the information to help guide our vision, but light can be seen with our eyes closed and light can be defined as a whole lot more.

- If electrons are pushed away from their nucleus, the collapse of the electron becomes light. We call it a laser. Because <u>light is motion</u>.

- Compressing a crystalline lattice produces light and electricity. This is called Piezo-electric effect because <u>light is pressure</u>.

- Electromagnetic waves disrupt and modify our brains when the vibrating frequencies are low because <u>light is a disrupter</u>.

- When the frequencies are very high the light can kill a person because <u>light is dangerous to matter</u>.

- Changing the frequency changes the perceived color, whatever color is because <u>light brings life to our world</u>.

- Photons seem to go faster than the speed of light which means they would be going back in time because <u>light travels between universes</u>.

- At the speed of light mass goes to infinity, but photons seem to have NO MASS because <u>light doesn't exist.</u>

- Photons are nothing more than a vibrating quasi mass; whatever that is. These <u>quasi-mass things are not light</u>.

- Photonic Force is one of 3 dimensional controls that allow motion in our universe as <u>light is the perpendicular representation of life.</u>

- <u>Light seems to define or represent quanta</u>.

OK!! That didn't go very well, so let's continue.

Life Is Not Life

While I know you are getting confused, looking at photons is the easy part. Understanding life really will need a good understanding of light as it has a similarity without the free will, so to speak. As a backdrop of our life discussion, let's see how time affects our lives. Some just think they go through life at one rate and then they die. Others may understand that life is simply a vibrational dimension, but they have little idea what that means. I went through the "transverse viewing" analogy and showed how, in absolute time, someone experiencing transverse viewing can see all of time in an instant including what will happen in the future. Besides that; in the last book, I talked about reaching out to our "linked" universe to travel backwards in time.

Let me try something a little different in describing life by using light. While light is constant, we have also determined that there is no aging if someone is going the speed of light to us. It makes no sense to us, but we sort of accept it. Atomic clocks being sent in space ships come home showing the "experienced time" in the space ship was reduced. From these experiments and others, there is little doubt that life is somehow suspended at high velocities.

Unfortunately, it is not a simple observation. If someone was going to the nearest star at very close to the speed of light, he would get there in about 4.5 years. We could watch the event and record the event, but the man in the ship would still not age. His universe slows down as he speeds up. To him almost no time passes. If he shines a flashlight about halfway to his destination, the light from the flashlight would get to the destination before he got there in his world, but about the same time for all other viewers.

This reduction in aging is the best evidence that the Life dimension is associated with individual universes or associated with universes linked by a common velocity. Remember it doesn't matter what direction you move to cause this affect. If you were simply vibrating at the close to the speed of light you would age very slowly and everything around you would simply start rotting before your eyes.

We can believe that because light goes the speed of light, it is suspended in time.--When I stated that Light was not life. There may be caveats to that. Anyone going the speed of light apparently becomes light and someone viewing transverse time possibly sees life as light and light as life.

Carnal and Spiritual Life

The second part of that statement doesn't work for normal no-mass light and carnal life, but some aspects of living are not carnal, they are spiritual, and sometimes light exhibits the effects of mass. I know it sounds like I won't have to expand on what life is after these statements, but I'm sure you would throw a book at me so I will continue to build a picture of light and life so that it will begin to make sense.

As a warning -Like the Surgeon General-For those not believing in a spiritual life, you will have problems with ANY true interpretation of Light or Life.

With that off my chest, let's look again at light and re-evaluate elements as we begin to show similarity and consistency with definitions we can present about life as well.

What Is Light?

Maybe we now know enough to approximate a definition of this mysterious and important element of our world. Light certainly can be described as lateral time, but its definition must contain more than this.

Window and the Way

It is both a window and a way to an adjacent universe. Generally speaking, no mass can make the journey as the mass goes to infinity. This means that the spirit is the component of life that can do the journey as it has no mass.

Energy Transfer Medium

Light also is the transfer medium of energy from an adjacent world into our own. I not only establish what we see but it also pushes energy into atoms to hold them together. The method for establishing the energy is not known and is considered to be God.

Light is Lateral Time

God views our universe transversely as he views the beginning and ending of lifetimes instantaneously across lateral time or light. Life and matter cannot breach the speed of light boundary because mass cannot be established nor can life.

Dimensional Adhesive

Light is vibrated by Electro-magnetic, nuclear, and spiritual elements. But it is not electromagnetic. Light is generated in the conscious minds of people. We cannot understand light until we understand people.

Electromagnetic vibrations hit our eyes and begin modifying the matter associated with the chemical make-up of our eyes. Somehow----something amazing happens. Our conscious minds build a world of something we call light.

With that basic insight of light, we possibly can begin understanding what "life" is. After all; if we want to understand life we had better get cracking. Here goes!!

What Is Life?

I think you have probable wondered about this at one time or another and you have never gotten a straight answer. I know you have heard about someone climbing to the top of a Tibetan mountain to talk to some Guru about this subject only to find that his answer was no answer at all. Well! I'm going to try to relieve some of that mystery and flim-flam.

The answer won't be simple nor will it be straight forward. Before you can see the truth, some relearning is in order. Life is not an embryo expanding to become an animal or human and it is not the combination of DNA structure to form a plant of germ. Life isn't controlled by particles or electro-magnetism. We had to establish an entirely new set of dimensional qualities to even attempt to look at life. Unfortunately for those who wish to see our universe devoid of a Creator and see it as a machine continually building, destroying and building again, you are going to have problems. There are some who try to define life with chemicals and sperm, but the ancient book of "Secrets" found among the Dead Sea Scrolls may help us in our desire to determine the truth.

Book of Secrets

The "Book of Secrets" found with the Dead Sea Scrolls provides a strong warning about the use of "secrets of God". The main secret implied was the secret of life. It says that because we, as humans, don't understand what we are doing as we manipulate "Nature" it will only end badly if we try to build "Life". Of the secret elements indicated in the text, it seems that the "manipulation and attempts at creation of life" is the worst on to try. Here is a small excerpt of the book. Its message is repeated over and over in many of the ancient texts. (Portions in parentheses were unintelligible.)

"--With (this I beseech your attention. All of the) secrets of sin (and life were attempted) but they **[preflood humans]** did not know the secret of the way things are nor did they understand the things of old. They did not know what would come upon them, so they did not rescue themselves **[as smart as the preflood people were most died anyway.]** without the secret of the way things are. ----(God controls) every secret, and he limits every deed and what (magic that is known by) the Gentiles **[People that were not descended in the line of Noah]**, for He created them and their deeds also ---You have *not become wise in understanding (my secrets); for you have not properly understood the origin of Wisdom."*

The book is saying that God warned us against trying to create life.---It can't be done because no one knows what life is. The Sumerian texts tell us that ancient humans "created" huge animals to fight against God in the ancient wars. Jewish, Greek, Egyptian, and Sumerian texts talk about misshapen people and animals that were the results of experiments with life. The Bible indicates that most of the animals that live today were ABOMINATIONS. We certainly can assume that these abominations were attempts to create life by modification of genetics, and other inappropriate things. If we try to understand life by trying to manipulate DNA and build new species, we will suffer for the blatant disregard for God's warnings. That being said, I don't think the quest for understanding what life is should be considered in this warning. After all, God made us inquisitive.

A new type of physics must be introduced to you and reality as you know it needs to be completely redefined. The beginnings of our journey takes us again to Einstein and his observations of reality and how very odd it is, but then we must step back and look deeper into the 10-dimensional universe to actually sense what it is that makes life almost impossible to define. In a 3-dimensional world definition is not only difficulty, it is totally impossible. String theorists tell us that there are at least 10-dimensions but the theories didn't know what to do with all the dimensions so they talked about them as useless mathematical anomalies. Well! Hopefully you understand by now that they are not anomalies and here is the kicker.

The ONLY way to define Life is by looking beyond the incorrect, generally limited, unfounded, and carelessly defined 3-dimensions we have been taught in school.

Let's look at some of the many problems that keep popping up.

Problems

There is no way I'm going to list the hundreds of issues encountered every day with 3-dimensional space. This book starts off with a few of the problems and it finds solutions along the way. It won't be easy, but I think you this book and the details contained here will be most useful to your understanding yourself and every living thing in any universe.

- **Problem 1**: According to Einstein, vibrational nothingness associated with mass and light continues outward from a central point and escape the limits of the universe at infinity. This means that sooner or later all will be dark and no mass will exist. While that would certainly be in the distant future, it still makes me uncomfortable and it really doesn't make sense that existence is not renewing. Everything else in the universe renews and we can, pretty much, be assured that the same is true of mass and light.

- **Problems 2**: The idea that time runs a similar path to mass and light and is a one-way adventure through our universe until one day it simply ends makes no sense. Time must be rejuvenative. This time thing is just as mysterious as life, but, hopefully, our discussions about light are beginning to open your eyes.

- **Problem 3:** Women have more genes or nucleotide bases on their chromosome pairs than men do. Men have 44 autosomes but only and X and a much shorter Y chromosome, while women have 2 X's. The average number of nucleotide bases [**A**denine, **G**uanine, **T**hymine, and Cytosine] in a male set of 46 chromosomes is 20 thousand. The super race known as women have more. Being a male, this concerns me as well. Should I simply worship the more gene-ated women and let them take control as they should or is life made up of something far more esoteric than Guanine. After all, a dead chromosome and live one look the same and have the same characteristics. Something outside the chromosome makes it alive.

- This thing we can call life or "self" seems to also be "Conserved" in this universe. As one set of chromosomes becomes dead, another seems to become alive. I know you are thinking there are more people alive today than during the old days, so this Conservation of Life" doesn't look like it will hold water. Please hold your criticism until the end of the book I think I can turn you around before that time.

- **Problem of the Problems**- What we actually will find out is that very little is known about what mass, time, life and many other components of our universe are. Please, please do not get frustrated with this book. It will truly help you understand life, living, God, Heaven, Time, and the endless mass talked about by Einstein.

Fake Mass

Before we get much further, let me first make a statement that I have presented several times now. There is no such thing as mass as we think of it. Don't get me wrong. Mass certainly exists. We see it, feel it, and smell it, but new revelations on how the component parts of our universe interact have opened our eyes. The entire world and everything you smell, touch, or see, is actually nothing more that nodal displacements of vibrational pulses that are, sort of, the heart of the universe. We will look at 10 major vibrational elements or dimensions that make up the universe and show how they must be held by strict rules of symmetry. What I mean by that is the only way to get more mass is to lose mass from somewhere else. If we want to experience more forward time, some place must experience the same amount of negative time.

Here is the kicker. Life like most of the other things you thought you knew about are only vibrations of nothingness. Don't put the book down thinking this is the answer that will be finally established. That is a cop out and we need to understand a whole lot of things that will seem foreign to you before we can get a better answer.

Universal Symmetry

Instead of using the theory of conservation of energy, which everyone seems to latch on to, in this book, we must look at a broader viewpoint. This is an expansion of what is commonly known as Super Symmetry which states that if mass gets small in this universe, it gets large in an adjacent one. It's a little more complex than that, but the main thing is that for our universe to work, we must be indelibly linked with another "Symmetric Universe".

Super-Universe

Today scientists are realizing that this symmetric Universe is REQUIRED for our universe to exist. So much so that one could just describe this universe dual as a super universe. In the past, there was a huge struggle to make our universe work by itself. No one could truly describe matter, time, motion, energy, or just about anything else. Let's review and expand on some of the pain that has taken place.

Baryons and Bosons

For some time, atoms were considered the building blocks of matter, the problem was that there were so many types of atoms and new ones were being created every day. They became a useless standard. Baryons like protons and electrons came to the rescue for a while. Baryons did work out so well either.

Einstein and others determined that between each Baryon there was an unseen component. Not only was it unseen, it had no mass. Einstein took the name given by John Keely in the 1880s and called this component "Aether". Whatever made up Aether could be the unified particle that made up the universe.

He knew electrons weren't the smallest division of an atom and the smallest component of life so he searched for this "unified particle thing" that everything was made from. He never succeeded in finding this thing, because the essence of the universe is not a particulate. If you have heard about the unified particle theory, this section cannot be aided by that. Einstein spent the last half of his life searching for this particle that made Aether and failed partly because he tried to hold on to a single universe association.

Einstein had spent his whole life searching for something that could not be. Along the way he determined that life affected things around him. In fact, consciousness was required for anything "normal" to exist at all.

What in the world are we to do????? If you could see me now, you would see that I want to tell you, but I have to ease you into it so you don't start thinking the whole concept is nuts. Oh well! Here goes a big part.

Eureka!

The answer to life is both simple and almost impossible to understand. Everything---I mean everything [including life], is made of vibration. I know you know about the problem. Here, vibration means vibration of nothing defined into a space that is controlled by a vibrational time that is coregent with a backward time component of our adjacent universe. This vibration concept includes all matter, all electro-magnetics, all nuclear energy, all photons, and all life forces. In studying the details, we may get a closer picture to what is wrong in Einstein's equations and we will get a better picture about how dimensions of the universe recycle. Their existence is "renewed" in the universe or the universe would, one day cease to exist. One might consider this universal conservation of sorts.

Universe Conversion

Here is a problem that has made universe investigation difficult in the past. We think of the universe as being confined to rectangular coordinate dimensions. In order to truly study it, we need to convert the universe into a vibrational concept. When that happens, we find several things. First, we find that conversion of and control of matter is no longer a mystery, it is reasonable and the concept of heaven is no longer impossible. Heaven now becomes necessary as it re-supplies our universe as our universe re-supplies the Heaven universe. The concept of transference to another universe, say heaven, is no longer some figment of the imagination of a religious leader. It now becomes a true probability. This is a concept that will help you understand yourself, your religion, your life and death, your entire universe.

Just to reiterate what I'm trying to say and show you in this set of books is that the smallest component of matter isn't matter at all. Further I'm trying to say our universe is actually a dual universe. If the 2 universes separate, bad things will happen. Lastly, Am saying life is vibrating nothingness.

Many of the present theorists know that this is the only self-sustaining theory that has not been smashed into bits. String theories, black holes, big bang, super symmetry, Tamashii modeling of mass, and all the rest can't be defined completely without external universes. That is because all dimensional qualities of this "seen universe" are in direct opposition or resonate with a collocated, unseen universe some call Heaven, some call universe B, some call linked universe. The thing that holds the universes or the second half of this universe is vibration.

Life Conversion

It seems that everything converts to other useable characteristics in our linked universes. As out-waves push their way to the limits of our universe, they are characteristically converted to in-waves used to apply stresses on our linked universe. As out-waves emanate from our linked universe into ours, they are converted to in-waves to apply forces to our out-waves. That whole machine, while exotic, sort of makes sense by now, I hope. I know the quantum mechanics dilemma threw some of you as time space is not a major definer of vibration, but that was mostly because our concept of vibration is through space and that is not what it is, exactly.

142

More importantly, life must have the same conversive effect in our joined universes. It's bigger than being born, living and dying.

It's more expansive than "I think; therefore, I am!" It is more noteworthy than "Life is a box of chocolates." It is more complex that "I die and go to heaven." It's more complex than light and it's more wonderful than the birth of a new baby. Before we can continue into the meaning of life, I think you need another little break again. Last time I simply let you rest on your own. This time we will have a short visit with lizards.

Not About Lizard People

Certainly, I said that the study of life and the study of death will be adventurous, but some things we simply cannot go though in this book. For instance, it seems like every ancient society wrote about, described, worshiped, and sculpted images of lizard people. I know I promised a discussion about what life and death was, but this would be way out of line. The ancient texts talk about these people and we can be pretty sure these unusual people live with the more normal people at one time, but that is a totally different subject. Ancient texts said that they were not only reptilian but also very ugly. The images following are a very small fraction of the drawings and statues found so far from around the world. There is no doubt that these were not "Normal" people. One could say this oddity is similar to the Einstein $E=MC^2$ equation oddity. The derivation is completely different than all the other energy equations used by everyone to describe our world. While I will discuss Einstein, I'm not going to talk about this particular oddity.

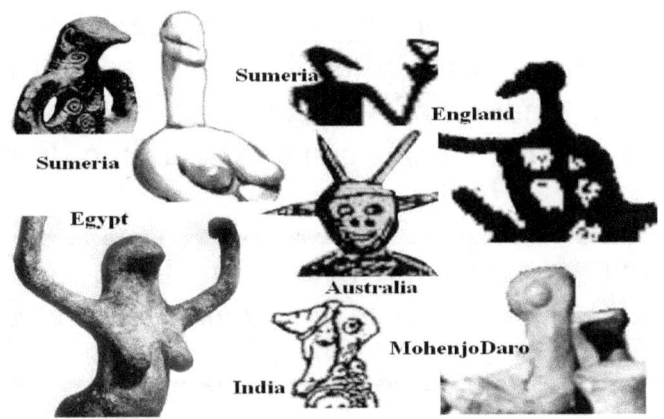

While many of the people simply had lizard like heads, some of the images looked scary as shown below, but these guys evidently lived with people for a time and apparently, they held honorable places in society because people got comfortable with their oddness. For a time, we could be comfortable with atomic theory and a 3 dimensional world, but all that is changing.

Lizard people are gone and we now must look at the world differently. I wrote a book about these lizard people, but now we must concentrate on life and the even more difficult question, "What is death?" Possibly a way to start is to look at some of the findings of Immanuel Kant and going back to lateral time viewing. This is going to be a little strange but it sort-of ties light and life together.

If you went the speed of light how long would you live?----
Your Right!!!--- Forever in the stationary universe is the
correct answer.

If you lived forever, you could experience many lifetimes in the stationary world----I know that sounds bizarre, but humor me a little.

Seeing The Future

In the previous book of this series, I dealt with time travel and all that stuff, but all of these strange phenomena are connected together in one seamless existence so we must recognize that there is a connection with light. In a speed of light perspective, man could go just about anywhere and quite a number of people, apparently, have done just that. Additionally, it seems that we can experience several predecessor and future lives that are unveiled during the process of hypnosis therapy. This is also well documented. To, possibly, help explain these things a little more, let me revert back to an old philosopher Immanuel Kant and several other similar teachers. They surmised that people could be represented as hanging entities in an empty environment.

Don't worry about this example. It is just another way of representing lateral time, but it might give you a better viewpoint.

The people [everyone and everything] might be hanging from a floor or ceiling as shown next.

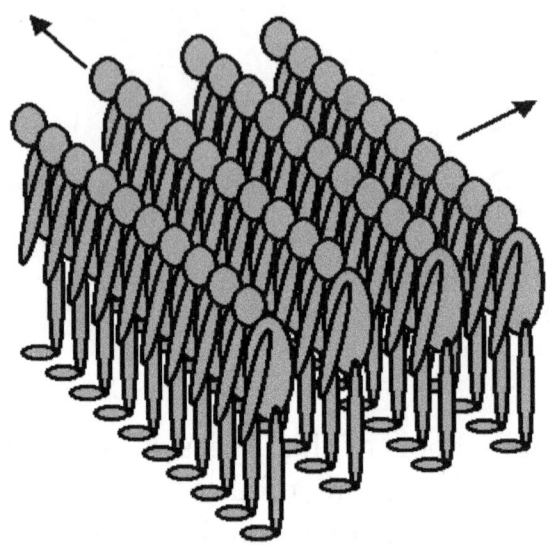

God would represent a reality to individuals such that they would live in a world outlined by the details put into their consciousness. This is represented in the following diagram.

God

Generally speaking, 2 or more people would have similar inputs and they would react to the "introduced" environment. [While this seems very odd, it seems to come close to Einstein's theory that we influence or produce our reality.]

Now that I have gotten you confused, let's say God introduced several environments into a person simultaneously. One might be in 10BC; one might be in the 15th century; and one might even be in the future all at the same time to the viewer of lateral time. The consciousness of the person might be able to remember all the events because he was a part of all the events. He could "remember" the future. The problem is that in this philosophical world, nothing is required to be actually real.

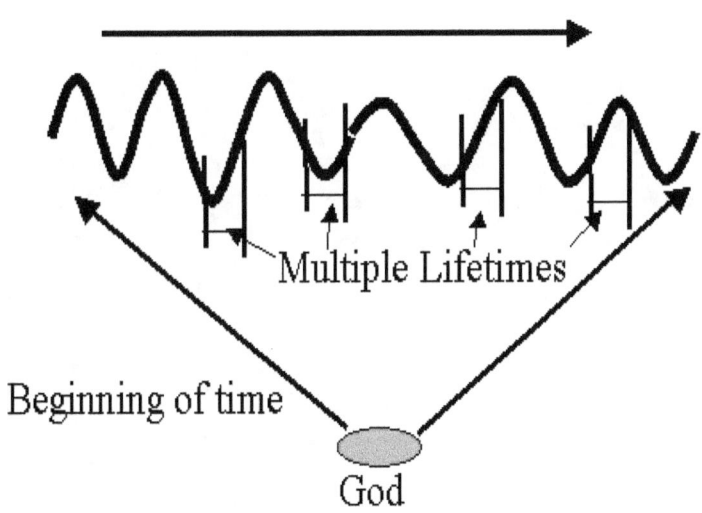

Let's Make a Reality

While this lateral viewing thing seems to work with God seeing the beginning of time and all, what if we could view the same lateral timeline and what if each one of the different lifetimes that are viewed are different people or the consciousness of the same person being "recycled" over time? What I mean here is what we typically call reincarnation. If one could interpret lateral timing, one would be able to see the future and the past as if it were all accomplished instantly. Just think about it! Rather than the perceived body being the essence of life, the essence would be defined in the speed of light world. Let me continue with I know I sort of getting into "death" right now, because you can't have multiple lives without multiple deaths so let me back up and push the life part.

Edgar Cayce on Light and Life

Edgar Cayce, the sleeping prophet of the twentieth century may give us a little more perspective.

"In the manifestation of all power, force, motion, <u>vibration</u>, that which impels, that which detracts, is in its essence of one force, one source, in its elemental form. As to what has been done or accomplished by or through the activity of entities that have been delegated powers in activity, is another story. [Of course, the one source he is talking about here is VIBRATIONs not recognized in this universe.]

"God, the first cause, the first principle, the first movement, IS! That's the beginning! That is, that was, that ever will be! The following of those sources, forces, activities that are in accord with the Creative Force or first cause - its laws, then - is to be one with the source, or equal with yet separate from that first cause. When, then, may man - as an element, an entity, a separate being manifested in material life and form - be aware or conscious of the moving of that first cause within his environ? Or, taking man in his present position or consciousness, how or when may he be aware of that first cause moving within his realm of consciousness?" [Cayce is indicating here that the first because vibration is the thing we must be conscious of if we are to understand life.]

"In the beginning there was the force of attraction and the force that repelled. [Milo Wolff would have said IN-waves leave the universe and Out-waves enter the universe.]

Hence, in man's consciousness he becomes aware of what is known as the atomic or cellular form of movement about which there becomes nebulous activity. And this is the lowest form (as man would designate) that's in active forces in his experience. [This is the basic Carnal Life haplessly enjoyed by most.]

Yet this very movement that separates the forces in atomic influence is the first cause, or the manifestation of that called God in the material plane! Then, as it gathers of positive-negative forces in their activity, whether it be of one element or realm or another, it becomes magnified in its force or sources through the universe. Hence we find worlds, suns, stars, nebulae, and whole solar systems, moving from a first cause. [These would be a parallel universe filled with the same things we find in the "Carnally linked world". The description of God being the first cause speaks to the other universe linked to spiritual life.]

"When this first cause comes into man's experience in the present realm he becomes confused, in that he appears to have an influence upon this force or power in directing same. Certainly! Much, though, in the manner as the reflection of light in a mirror. For, it is only reflected force that man may have upon those forces that show themselves in the activities, in whatever realm into which man may be delving in the moment - whether of the nebulae, the gaseous, or the elements that have gathered together in their activity throughout that man has chosen to call time or space. And

152

becomes, in its very movement, of that of which the first cause takes thought in a finite existence or consciousness. [Finite existence is the carnal element of the Universe and its negative time characteristics are certainly described as a mirror or mirror image of our universe.]

"Hence, as man himself applies himself - or uses that of which he becomes conscious in the realm of activity, and gives or places the credit (as would be called) in man's consciousness in the correct sphere or realm he becomes conscious of that union of force with the infinite with the finite force. [This is speaking to the possibility of changing one's "resonance" by acceptance of the spiritual universe and its implications. Finite is carnal, infinite is the infinite spiritual life.]

"Hence, in the fruits of that - as is given oft, as the fruits of the spirit - does man become aware of the infinite penetrating, or interpenetrating the activities of all forces of matter, or that which is a manifestation of the realm of the infinite into finite - and the finite becomes conscious of same. [Going from the realm of the finite to the realm of the infinite speaks of becoming a spirit through change in resonance or complete separation from the Carnal body in death.]

"It may be said that, as the man makes in self - through the ability given for man in his activity in a material plane - the will - one with the laws of creative influence, we begin with: Like begets like - As he sows, so shall he reap - As the "man thinks in the heart, so is he." [As Jesus said faith of a grain of mustered seed could allow moving mountains with our minds.]

"These are all but trite sayings to most of us, even to the thinking man; but should the mind of an individual (the finite mind) turn within his own being for the law pertaining to these trite sayings, until the understanding arises, then there is the consciousness in the finite of the infinite moving upon and in the inner self. Life in all its force begins in the earth as the moving of the infinite upon the negative force or the finite in the material, or to become a manifested force." [We can stay in a carnal world or enjoy the fruits of that experience that is beyond.]

To Dr. Casey and the insights that he obtained from his visions, or whatever they were, life was an extension of existence. Not only could life modify existence, life was and is existence. Without consciousness, Einstein told us there was no reason to define existence. It goes farther than the "Positive Thinking" described in a number of books where believing that you are rich and miraculously you will become rich. It is more like the entire universe you perceive is slightly different than all other universes perceived by other people because they are defined by one's life resonance.

A Conscious person not only can control his destiny, but also manipulate the world perceived by others around him or her.

Welcome to the Ethereal Dimensions, Adjacent universes, and our Ekpyrotic membrane.

Ekpyrotic Membrane

Matter Seed

The Ekpyrotic /Membrane Theory is a good one. It indicates that there must have been at least **2 adjoined universes** that splatted together to initiate what we call the Big Bang some 15 billion years ago or the Big Bang simply could not have occurred. This SPLAT action caused everything to emerge from a 'fireball' with a temperature of 10 billion degrees. When the two splatted together, the energy that became matter was introduced. According to this theory tiny quasi-particles called fermions were all over the place. If you are trying understand what a fermion is, forget it. It is a nothing that sort of becomes something when an action is applied. In my book on the 10 dimensions of the universe, I simply identify fermionic force as one of the dimensions needed to produce a universe. If you think of it as a dimension, you might have better luck. Another possibility is to think of a fermion as a matter "seed".

When the universes collided 15 billion years ago, the fermions got all twisted around and [by magic] turned into complete particles we call Bosons. It is these bosons that make up most of what we call matter. The question is what actually caused the fermions to become matter and what things are necessary in this universe for sustainment?

The reason the fermions became matter is simple. They began to vibrate.

With that, I eliminated the fermion, quasi-particle for something to be called a vibrational node in previous books and we should understand, by now that modern day theories REQUIRE more than one universe. As far as what an adjacent universe I've been calling it heaven. There certainly could be many more adjacent universes, but Ekpyrotic membrane required at least two. For a practical observation concerning these not-quite-a-particle things are, we need to look at light closely; especially when we are investigating life. I don't think you understand how extremely odd it is that our "so much less than the speed of light" lives are so greatly affected by the fast paced "going the speed of light" photons are. One would think they can be in the same existence or at least one would not be able to interface with the other. For there to be continuation of light or anything else, we must establish some more Theories.

Conservation of Everything

Here is the main problem. We now know that everything is vibrating from energy nodes [or fermions]. The vibrations NEVER stop. They continue to move away from the central core. Just as Einstein had theorized, soon, they would be lost at the end of the universe [whatever that is.] We would run out of particles, photons, and energy. **We would run out of life itself.** Luckily, we find that there is a conservation of everything that sustains the universe.

Conserve Energy-It's the Law

Let's first consider the common "law" of conservation of energy. All we can do is to change the type of energy we cannot add to it or take away from it. In recent years some have identified areas that seem to create matter which suggests the creation of energy. If we create matter, certainly we create energy and we create something else that is very important. What we will find out that conservation of energy is not a true concept, but symmetry of energy absolutely works and I'll show how it affects life. Not only is there symmetry of energy defined in our universe, but also the following:

- **Symmetry of Energy** While Einstein indicated that it could not, by itself be conserved, we find that by a new theory called Super Symmetry; energy can APPEAR to be conserved.

- **Symmetry of Matter**-As a black hole produces matter, there is no issue as anti-matter is reduced at the same time. I'll have to make anti-matter more than a comical word for you on this one.

- **Symmetry of light**- Einstein was always worried that as these photon things approached the end of universe soon, the light would leave our universe and the number of photons would continuously be reduced. We will see that Light is sort of rejuvenated by anti-photons, but that by itself does not allow us to appreciate the electromagnetic vibrations. That required a little Ethereal physics.

- **Symmetry of life**- As people die others are born. As people become more spiritual, other become less spiritual. If someone's life is ended here, it is started in another universe and vice-versa. I know this sounds like yen and yang but it is more than good and perceived evil.

- **Symmetry of Death**- As people die they must be born or their consciousness does not continue. If it does not continue, the reality established by that consciousness ceases to exist and reality will become more and more skewed. Soon physical laws would have to be changed to support new life that had not been established from a previous life. While this is not a true symmetric requirement, we will find later that death must be recycled just like life.

Everything in the universe is regenerative. We simply need to broaden our horizon to find what is feeding the dimensional elements of the universe

This one is part of the answer to that Life Question. The thing that is continually created is time.

We must be creating "TIME" as fast as we lose it.

Time Re-creation

I know all of this sounds bizarre so I'm going to stop this whole thought pattern [AGAIN] and sort of go through the history of how we need to identify what Mass, Energy and Time really are and we will have to bring up another thing as well.

Operational Dimensions

One thing we cannot ignore right now is that a universe with only height, width, and depth cannot produce energy, there must be contention. What we will find is that the universe is critically designed to have equilibrium of vibration. Those emanating away form initiating points are the ones that make mass, but vibrations [called In-waves] are moving towards the universe center just as fast. These vibrations come from --------outside our universe. I know I have been pounding this into you, but this is a very important component that needs reinforcement. Even with that, it does not directly answer the Life question. To get closer, we need to expand the set of dimensional elements known as the ethereal dimensions. It just so happens that these dimensions all are REQUIRED and displayed by something we call life. I know you are saying Bah-humbug. Just quit it and you will see the beauty and final simplicity of all of this AND you will find out what life truly is. Part of it can be witnessed as a conscious collective.

160

Conscious Collective

Another way of looking at this ethereal component of the universe has been ADRESSED is to call it the conscious collective, but what one really needs to understand to fill in some of the blanks was presented by Einstein many years ago now. That idea was that relativity is not simply how space and time are modified to present a common image to an observer, but that –

EVERYTHING seems to be modified DIRECTLY by this conscious collective thing-a-mo-bob. Don't go saying "We are Borg" like the Star Trek Nemesis, but in a way, we establish a unified reality.

I'm going to get you comfortable with this whole ethereal dimensional model and the rest is up to you. Once you understand how it all gets put together, you will have to build on these kernels.

.

General Awareness

Before I get away from the weirdness, let me sort of present a dynamic that looks more and more like a reasonable depiction of our universe.

Everything works in similar ways.

On the surface, that doesn't sound so bizarre does it? But what this entails will be far from natural sounding.

If you did not read the book on the 10-dimensions of the universe, some of this book might seem odd. To build matter there are 3 dimensional elements that are mutually perpendicular. By themselves, a component of the universe cannot be completed. The force producing in-waves have 3 dimensions that are mutually perpendicular and these are also needed before atoms can be atoms and planets can be planets. With all that being said, general existence of things to us cannot be accomplished without 3 more dimensions of consciousness.

Einstein indicated that the Consciousness seems to control the universal concept of time and space.

In my presentation I called this third group of mutually perpendicular dimensions the Ethereal Dynamo. The last of the dimensions we can understand is the thing we represent as time. Time pulls together the three-dimensional dynamos in a mutually perpendicular way, together.

Because time, in our minds is a reference of vibration, time is, sort of, integral in all dimensional components as all 9 dimensions building the universe are made up of vibrational nodes.

OK! Enough review. Let's come up for air for a minute before we go on. Are you OK now? Please try to hold it together because I am really trying to help you see a new concept in matter and the universe rather than trying to make you think I'm one of those guru, head-in-the-sky, ding-dongs that answer questions with undefined answers. Let me review and expand just a little more so we have the proper backdrop for understanding life a little bit better.

1. **Matter ain't matter**. Instead, it is made up of what Einstein described as undulating nothingness. I will call it vibrations. [The same can be said about what I call the operational dynamo made up of these in-wave things to make forces. The Ethereal dynamo is made up of these vibration things as well. Everything has a similarity and nothing has an end.

2. **Nothing has an End**-The important thing that many now profess including Einstein and Dr. Wolff is that matter never ends. It sort-of fades away as the vibration ripples get father and farther apart away from this seed, sometimes called standing waves as the in and out-waves seem to cancel each other out at one point in their travels. It is this neutral point that could be considered Fermionic.

3. **Vibrations Must be Replenished**-As vibrations are lost in out universe, they must be replenished from another

universe. Later we will see that it is the nature of time in an adjacent universe that actually allows for sustainment, but right now just know that there is conservation of everything so anything leaving must be "somehow" coming back.

4. **All Dimensional Components Vibrate**-These matter vibrations are moving in 3 mutually. perpendicular ways. All dimensions that make up the universe vibrate including dimensions in Structural Dynamo [at make up stagnate matter], the Operational Dynamo [that make forces] and Ethereal dynamos [that produce conscious life]. Everything is similar.

5. **All Dimensions are grouped in sets of 3 perpendicular components**-The Operation dynamo is the one we know the most about, because it causes motion. In our infinite wisdom we called one of the dimensions electric fields which we know to be perpendicular to another dimension we typically call magnetism which is mutually perpendicular to the third of the operational dimensions that sort of joins the characteristics of both to produce Electromagnetism and photons. Guess what!! The other 2 dynamos must work the same way.

6. **Stagnate Matter is three dimensional**-As an example we know that anytime particles are produced, all of a sudden there is something we call gravity. Putting a vibrating fermion and gravity together is something we can loosely call Nuclear attraction. All three of these things produce what we might think of as stagnant matter.

7. **Life is three dimensional**-As another example, no one has a good definition of life. It certainly is not DNA, but

164

most of the ancient religious data tells us that a being is made up of three entities [consciousness/self, soul, and spirit] or [id, ego, Baa] or on and on we could go. These three ethereal dimensions act just like the other 2.

8. **All Dimensional groups resonate**-We know that any electric and magnetic field have a particular vibrational quality that places them in something we call **resonance**. Resonance is when dimensional qualities place the least or the most stress on the universal structure. Dimensional duals try to vibrate at this resonance as it provides the most stability. Guess what! The other 2 must have this same characteristic.

This is important ---LIFE RESONATES.

---

9. **Dimensions take in and give off energy out of phase**- There is a certain capacitance and induction that is noted by 2 dimensions in a dynamo. While we know that to be a fact in the operational dynamo with electricity and magnetism and that both characterizations are phase shifted from the other so that it is not sensed by the other. ---Any dynamo should have the same characteristic. Capacity or capacitance is the capability to take in or secure energy and induction is the ability to provide or use energy. Because the dimensions are mutually perpendicular, the 2 linked dimensions of a dimension characterized by a capacitance cannot sense the capacitance just like Induction is hidden from the others.

10. **All dimensional Energy Equations MUST be able to be presented in the same form**.-If one energy equation is represented by one dynamo against time. All energy

165

formulae should be represented in that same way. This last one is the issue that I have with the $E=MC^2$. This happens to describe energy or in-ways with respect to one of the dimensions in the Particle dynamo and other formulae have a different structure. $E=1/2\ CV^2$ of $E=1/2\ LI^2$ which describe out-waves with respect to dimensions of the Operational dynamo.

11. **Time is a dynamomic dimensional component that ties the three dynamos together in a mutually perpendicular way**. - We can look at time as another dynamo made up of the three dimensional dynamos. It is sort of a super dimensional dynamo. Without any one of the other three, time would not be possible.

12. **Aether, Electricity, and Carnal or base Life** are joined together as static charge or building characteristics of their associated dimensional dynamos.

13. **Magnetism, Gravity, and human spirit** are joined together as the kinetic elements of the dimensional dynamos.

Dimensional Dynamo Chart

The Chart below comes from the second book of this series. Form it you can see the relationship Time has to all three-dimensional dynamos.

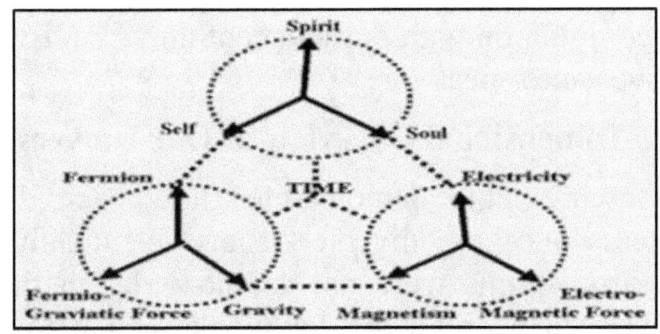

The Dynamos are held together by the interaction of in-waves and out-waves, but just what is an in-wave and is there a corollaric in-wave for life??

In order for life to exist and continue, there must be the rejuvenation of life that comes as in-waves from outside our universe.

God Took the Tree of Knowledge From Man

Life combines self, spirit, and soul to the universe in a special and important way. Once we see the significance of our particular lives, we can begin to understand how much power we possess in this universe. There was a reason why God pulled the Tree of Knowledge away from people. Whether the description was figurative or descriptive, there is one truth that is presented in the Book of Genesis. If man knew the innate power presented in his creation of conscious life, we would mess up this world and limit interaction with our creator. On the other hand, knowledge can be split into two confines—carnal knowledge and spiritual knowledge. As one expands the carnal nature, there is a loss in "communication" with our adjacent universal duals. Expansion in a spiritual way, changes our resonance which modifies the resonance of the perceived world towards a unified communion with our adjacent universe. Everything is interactive, so to speak.

Dimensional Overview of Our Universe

The following table demonstrates how each dimensional component of the universe reacts identically to their counterparts. If this were not so, no order in the universe would be seen. Don't worry about the actual formulae and component lettering. The main thing is that consistency of energy AUTOMATICALLY means consistency of the other dimensional components. While our emphasis in this section will be to the Ethereal, we must recognize that all dimensional dynamos are interrelated and interdependent.

Dimensional Dynamos	Operational	Structural	Ethereal	Time
Static Dimension	Electrical	Fermionic	Self-ity	Structural Dynamo
Kinetic Dimension	Magnetic	Gravitational	Soul-ity	Operation Dynamo
Tertiary[3rd] Dimension	Photonic	Nuclear	Spirituality	Ethereal Dynamo
Static Entity	Voltage [V]	Velocity [□]	Internal Life [□]	Reactive Time [t]
Kinetic Entity	Current [I]	Distance [x]	External Life[□]	Regenerative Time [□]
3rd Entity	E/M Wave	Atomic cloud	Life force	Time
Static Effect	Capacitive Force [C]	Mass [M]	Conscious force[□]	Existence [□]
kinetic Effect	Inductive Force [L]	Gravity Constant [K]	Life Constant [□]	Time Constant [□]
Tertiary Effect	E/M Force	nuclear Force	Consciousness force	Time Force
1st Reactance	$Z_{o1}=2\pi fC$	$Z_{s1}=2\pi fM$	$Z_{e1}=2\pi f\square$	$Z_{t1}=2\pi f\square$
2nd Reactance	$Z_{o2}=1/(2\pi fL)$	$Z_{s2}=1/(2\pi fK)$	$Z_{e2}=1/(2\pi f\square)$	$Z_{t2}=1/(2\pi f\square)$
3rd [Resonance]	$R_{o3}=(LC)^{-1/2}$	$R_{s3}=(MK)^{-1/2}$	$R_{e3}=(\square)^{-1/2}$	$R_{t3}=(\square)^{-1/2}$
Made from	In-waves	Out Waves	Reactive waves	Dynamomic Waves
Attribution/ conservation	light	Mass	Conscious	Time
1st Energy	$E_{o1}=\frac{1}{2}CV^2$	$E_{s1}=\frac{1}{2}M\square^2$	$E_{e1}=\frac{1}{2}\square^2$	$E_{t1}=\frac{1}{2}\square\, t^2$
2nd Energy	$E_{o2}=\frac{1}{2}LI^2$	$E_{s2}=\frac{1}{2}Kx^2$	$E_{e2}=\frac{1}{2}\square^2$	$E_{t2}=\frac{1}{2}\square^2$

Vibrational Distinction

As we saw, the Ekpyrotic membrane matter uses fermionic seeds as the "effect of collision" with 2 universes. I know that is bizarre, but these seeds, or nodes, vibrate. The faster they vibrate the denser the "apparent matter" becomes. As it slows, the matter appears to turn to light, but as shown below, what may really be happening is that the electromagnetic wave reactance in this world becomes stronger. In the operational dynamo, what we call light frequencies become radio waves as frequencies are decreased. In the diagram below the vibrational frequencies of matter are in the exahertz range [exahertz means "quintillion cycles per second"]. The frequencies have been converted to the distance traveled during one cycle at the speed of light so don't go out and try to whistle some gold in your pocket. Gold vibrates at 60 exahertz meaning it wavelength is 5×10^{-6} microns. That's really, really fast.

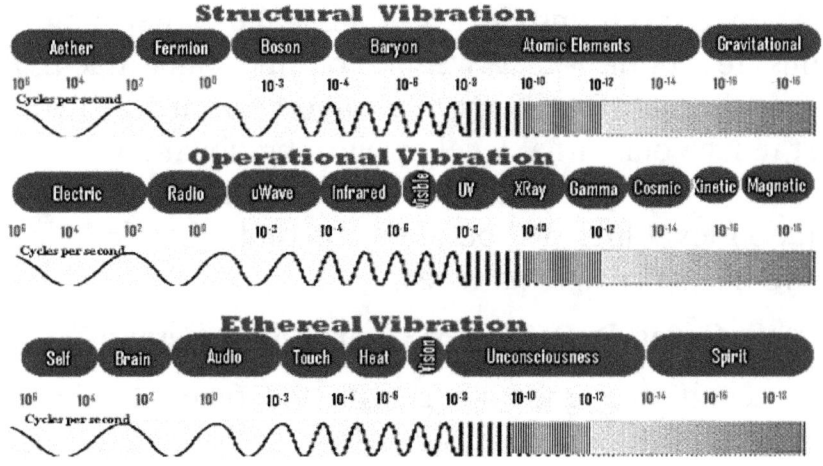

Quick Definition of Ethereal Spectrum

The main thing I want to introduce by this chart is the ethereal vibration spectrum. Notice that low frequencies can be described as self or carnal levels. As the vibrational aspect of this dynamo increases, the Spiritual component or sub-consciousness takes control and finally spiritual pureness is even at a higher level. Low levels are associated with survival, sex, and hunger. The lowest of these levels is life without true consciousness like a snail. Its life consists of survival with a very few ventures into sexual gratification and nothing more. One could say its life has potential, but no real force. The higher frequencies are associated with enlightenment and finally spiritual awareness. Guess what! If you become purely spiritual, you cease to exist in this world. All living entities stay somewhere in between. The noticeable characteristic of Life is that it has a resonance. What I mean by that is the more spiritual one becomes, the less carnal he must become. While this seems sort of obvious in a life description, the amazing thing is that the other 2

171

vibrational spectrums do EXACTLY the same thing. This begins to confirm that we are on the right track. Also understand that if we want to change our spiritual level, we must reduce our carnal level or our vibrational level must be increased. The question might be, "Just how can I get the vibration level higher?" before we get into that, let's look at electromagnetism and particles.

Quick Definition of Operation Spectrum

Let's see if that goes along with the other 2 vibrational spectrums of the universe. Let's look at the electromagnetic vibrational elements. At the lowest frequency level, just like the Ethereal spectrum, there is a vast potential of producing forces, but no actual force is produced until vibrational levels are increased and if frequencies continue higher and higher well beyond the cosmic radiation levels, the effect on the universe is completely magnetic. There is no substance at all. Between these levels is this thing I've been calling resonance. Resonance says that to be stable, if there is a high level of magnetism, there must be a lower concentration of Electricity and vice versa. Resonance in the electromagnetic dynamo has been studied for hundreds of years now so scientists know how things react. The only issue is that no one seems to be putting together the other two-dimensional dynamos and the similarity of all three in creating a oneness in our universe.

Quick Definition of Structural Spectrum

Let's look at the Structural vibrational elements. At the lowest frequency level, just like the ethereal spectrum, there is a vast potential of producing matter. Others call this potential the Aether, so who am I to change a name. It should be noted that no matter can be identified with Aether just like no real work can be associated with what we call voltage. Voltage is simply electromagnetism that could be used just like Aether is the essence of matter that could establish matter given the right nodal [being hit by those in-wave things] interfaces. No actual matter is produced until vibrational levels are increased and if frequencies continue higher and higher well beyond the heaviest elemental frequency levels, the effect on the universe is completely gravitational. There is no substance at all. Between these levels there is something called resonance. Resonance says that to be stable, if there is a high level of gravitation, there must be a lower concentration of Aether and vice versa. One way to look at this is simply this everything has the same number of vibrating things associated with it. On the fermion, Boson, Baryon, hydrogen side, there are more of these things that could be called Aether than on the Uranium side where many of these things have become true particles producing gravity. A black hole is close to being an absolute gravity and it generally does not exist in this universe. The reason I'm bringing all this up again is that life acts EXACTLY the same way.

Examples

Even though these concepts are really, really, strange, right now, hopefully, you are seeing the similarities of these three seemingly completely different components of our universe.

If you were me about now you would be shouting, "What about time in all of this?" Well, this even gets better. As I mentioned, time is indelibly linked to the three-dimensional dynamos. Each resonance frequency affects how time is established. In the case of time in is easier to see if we lump all matter, all life, and all electromagnetism together in huge lumps representing all the matter, force and life in the universe at one time. Each of these lumps has a particular resonance, just like every single entity. Time stays the same so long as all the combined groupings resonances stay at particular frequencies. This is where the relativity thing comes in. Let me just take life and move my perception point in the universe well away from the central mass of life [wherever that is]. Time will change, because "life would be resonating less carnally" at that point and this would shift the very nature of light, life, matter and forces of the universe.

If one got far enough away from carnal life, time would stop all together and finally it would go in reverse as time perception point was in an adjacent universe.

Fast Travel and Effect on Time

Let's take a different viewpoint. If we are traveling faster and faster, soon time will stop, according to Einstein and our size gets smaller and mass gets larger. Soon we would become a black hole as time ceased to exist for us. Unfortunately, we would be completely magnetism without electrical field and complete gravity without matter. Here is a weird part---assuming one could survive the transition, if someone went faster still time would go backward and the traveler would be the exact opposite completely electrical and Aetheric. One could assume it would be like antigravity and things would be pushed away. Where the traveler's mass had been large, it would now have to be nothing at all. Time has stopped, so what is registered in a person's life??

Life Becomes Light

In order to survive in this universe, the mass must have been shed. If mass is shed, a traveler would have become light. While in this state, there would be no conscious recognition as the operational dimensions of the entity would have taken control of the resonance. The man would have transformed to light during the travel to survive.

Light Reactions

As shown in the preceding diagrams, the make-up of electromagnetic waves can be defined as non-particulate matter that is, instead, made from in-waves. This characterization is pretty well known. Slow waves start as radio waves and graduate to light and finally terminate [as far as we can tell] as the killing Gamma Rays we all fear. We know that light with a wavelength of 0.5 microns is great stuff and allows us to see the world, but if it starts vibrating faster to 10^{-6} microns it turns into Gamma rays and can easily kill us. If a photon vibrates too quickly it becomes pure magnetism. If Photons vibrate too slowly they become Electrical potential or voltage. As we investigated earlier, the following must be noted.

Electromagnetic vibrations in the visible spectrum did not create light. That is all done in the mind as it is associated with life's vibrational level.

Just like we cannot see or sense in anyway pure electricity, we cannot sense the essence of matter or the essence of life.

Our concepts of light, matter, time, and life have to do with our state of being and our conscious resonance with in the universe. One can say that none of the 4 things actually exist.

What I mean is after all I discussed about light, it is only a conceived expression rather than an actual component. Sure, you need eyes to interpret various electromagnetic vibrations as color and a blind person must have a different perception of light due to his limitation, but it does not mean that blind people cannot see the light, nor does it mean that "sighted" people can see the light and life acts the same way.

Just like the others, life is regenerated by in-waves. While the electromagnetic dimension is the in-wave characterization of the universe, life is much subtler and more transferable as was discussed in the light speed traveler turning into light during his motion and then being reconstructed in the now time as the traveler slows. To understand life, we must understand in-waves. A question you might be asking about now is, "What is an in-wave of life??"

What Is A Life In-Wave?

I've been kicking around these vibration "standing waves/nodes" and these odd in and out waves, because some physicist identified them that way.

I'm sorry!!

Let's try to put a perspective on them. The out-waves are fairly easy to understand as vibrating nothings emanating from each of these standing wave point. OK! Not very easy, but at least one can generally understand the concept. Life out-waves are generated by conscious understanding of the environment. [Self, Survival, and Sex]. None of these attributes allow modification of the self-resonance and all can be identified as CARNAL or generated from inside the universe creating out-waves that spread out to the limits of the universe. [The body is inconsequential]. OK it's nice to have a body, but what I mean is that the Self, Survival, Sex life of a bug or germ or even a person doesn't affect the other dynamos. They are Static and generally identified as "the self-dimension". The out-waves keep going away from the "node" to infinity and --------beyond [never to return???????]. It would seem that life would vanish from the

178

universe over time.

Out-waves are leaving all the time. We need them back or our universe will collapse so in-waves come to the rescue or particles and life itself.

Life Doesn't Vanish

For life, out-waves constitute the carnal aberration of living. All aspects of self can be easily shown to have initiated in the universe and as they try to expand outward to our adjacent universe, then something magical happens. They are bombarded with in-waves of life just like particles were bombarded to establish stress in particles so that work can be done in the universe. This bombardment alone should allow us to know that other beings are resident in the adjacent universe. Their survival dimensional vibrations continuously leave and are converted to in-waves that CAN react with ours if we want them to.

Some More from Milo & Albert

To try and get a feeling about what in-waves are, Einstein and Dr. Milo Wolff will help. Dr. Wolff stated, *"Forces/Fields are caused by wave interactions of the Spherical In and Out Waves with other matter in the universe which change the location of the Wave-Center and which we 'see' as a 'force accelerating a particle."*

In English, this means that these in-waves make force to establish energy needed to run our universe. Unfortunately, there are a couple of real issues. Where do these all important in-waves come from and how do in-waves relate to living?????

179

In-waves come from outside our universe. Lucky for us, the outside universe senses time backward to us, so out-waves from our universe become in-waves and vice-versa. It's like a reciprocating machine set in motion and controlled by a creator.

These In-Waves Are Different

In-waves of living substance are somewhat different than those we sense as electro-magnetics. While their character and what they interact with are different. They should also be similar. As Carnal life spews out of this universe, we can assume it can be a driving force in our linked universe that has living substances spewing out Spiritual life that regenerates our Carnal living.

Life's Out-Waves

You may know what I'm going to say, but I will say it anyway. The in-waves are coming from an adjacent universe. When they come into our universe, they act backward to the out-waves produced as various out-waves come in contact with each other to produce vibrational nodes we describe as atoms or atomic clouds. Because the in-waves are backwards, contact with 'out-waves' cause STRESS we call conscience.

Think of it this way. Carnal out-waves stressing self/survival/sex emanate outward and they are countered by in-waves that stress the exact opposite which are true love, true concern or empathy, True willingness to die for another, true spiritual insight. The next time you hear a little voice telling you not to be selfish, or eat some food you should not have, or giving you some super strength to save a life, understand what that really is. It is simply the In-waves of life just like we have been going over and over as they are represented in the other dynamos. One can think of them as backward selfishness, or backward hate.

The in-waves are not backward to the adjacent universe. Time is reversed to assure conservation of time and life is reversed in the same way. In the adjacent world the Self dimension IS the Spiritual awareness/empathy, and love. The out-waves from our universe are mostly self/sex/and survival which are used by them to allow them to understand themselves in some way. If that isn't odd enough, as time goes forward here, the concept of time in our neighbor MUST BE backwards to us but forward to them.

Attack of the Etheric Vibrational Nodes

Because the in-waves are in opposition to all the in-waves, when they come in contact, they put stresses on the Etheric vibrational nodes. That is what we think of as carnal stress. Carnal stress has limited effect on our lives unless we can change our life resonance. That is coming up but it may start sounding religious so I'm just warning you now.

Life Would Disappear

If we had no life energy inserted from out adjacent universe, life would simply have no meaning at all. If someone tried to inform you of one basic truth about life it would be that living without a heaven is simply existing and life in our universe would eventually disappear. So, what! You created a great food dish or ran a mile, or saw a sunset, or became an evil dictator. The motions of life would simply have no meaning, they would be actions controlled simply by the environment. The word predisposition would be will rather than could be. There would be no moral or immoral. There would be no good or bad. There would be no "substantial" happiness or sadness. I say substantial here because humans and robots, for that matter, can fool themselves into thinking they were happy or sad or depressed or destitute. Possibly even a germ thinks he is having fun.

In the section on Light, I mentioned that the ancient Jewish

182

religious books and our Bible indicated that God took the light away from Satan and his followers and Jesus told his followers that he was the "light of the world". While there is some reason to classify these things with the study of light, another more reasonable way to view these statements is with respect to life. These statements could have been said a different way. God allowed the life resonance of Satan and his followers [From the adjacent universe] to remain Carnal giving them little reason to live.

Jesus To the Rescue

Besides coming to this universe and dying for our inability to accept anything from the spiritual in-waves, Jesus [incarnate GOD/Creator] brought a new way to expand our resonance and allow for spiritual in-waves to expand, rejuvenate and give meaning to our lives. Certainly, Jesus' resonance gift is not the only way to change one's life, but it seems to be a good one.

Oh Yes! Jesus' gift had a name [Holy Ghost].

I know it's a weird name for a gift that would allow us to modify our resonance and allow interaction with our adjacent universe while we are in a carnal body and even afterwards, but the name is just a name.

183

In and Out Wave
Differences

The reason out and in ways are opposite is that adjoining universes have "backward time" to each other. I know I've stated this a couple of times, but it needs to be understood. As an out-wave leave our universe, it appears to reverse by the time dimension of the outside universe and it is turned into an in-wave. Guess what!! The out-waves in that universe have vibrations in opposition with the newly created in-waves. As these out-waves leave the adjacent universe they enter our universe and appear to be opposite because of our backward time to the other universe. Of course, what that means is particles in this universe become forces in the adjacent universe. When studying life, one must also sense the difference in experience. Out-wave "carnal" experience is substantially different than "Spiritual [in-wave] experience. Both must be levied together to allow for a more meaningful life experience. God knew that Carnal existence and the feelings of self, sex, and survival were too strong so one of the three-dimensional components of GOD was sort of introduced to channel the spiritual in-waves into our carnal selves.

Not Sacrilege

I know some are thinking this talk is sacrilegious and others

184

are saying that I should not be confusing science with God and the third group is saying that this mumbo-jumbo means absolutely nothing.

Well, I see everyone's points, but the facts are as indicated below.

- God's trinity was described that way for a reason.

- The Holy Ghost being introduced into our world to allow us to "accept" a gift from God, sounds pretty much like what I have been saying and it is remarkably similar to the interactions of the Structural and Operational dynamos.

- There is always a struggle in ourselves to do the right thing followed by doing the most comfortable thing.

- There is an innate carnalness to mankind, no matter what we say.

- We know that consciousness does affect reality. It has been somewhat proven in experimentation and substantially in calculation.

- If life did not have a way to be rejuvenated, soon there would be no life in the universe just like there would be not light and all matter would be in a state of entropy.

- God being a trinity only makes sense if everything else in the universe is made the same way and dimensional characteristics are in groups of mutually perpendicular threes.

- The power of positive thinking and self-actualization absolutely do change our environment just as written about in dozens of books.

185

- Life is more than chromosome multiples in a body. That is simply stupid.

- Life is more than chemical reactions in a brain. Just like sight and vision are completely different, life and predispositions are completely different.

- The will to live is a very strong force that has nothing to do with the brain.

- The conflict of once thoughts to do good and/or something that will satisfy self is not caused by chemical reactions in the brain.

- When your body dies, life does not. It either stays in this universe or it goes to the adjacent universe.

- Some methods were discussed in the preceding book to allow travel into the adjacent universe to provide a mode for time travel.

Let's see how we can go to the next universe.

Going to Another Universe

Particle transfer

Let me take this one more step. Let's say a person from our adjacent universe goes faster than the speed of negative timed light in his universe. He would immediately be transferred to this universe and become force rather than matter. I know you are thinking about how watchers seem to be able to make this transition and can still be touched and are just like people so there is something else going on with life or the watchers wouldn't be showing up. This type of transfer could possibly be done without changing the resonance of life and development of a more spiritual lifestyle, but there would certainly be issues in understanding a world where their perceptions would be different that the perceptions of those around him. If a carnally minded person lived in the heaven universe, he would always feel completely out of place, out of sync, and unable to assimilate. When the ancient texts talk about how Satan and his followers had their LIGHT removed, possibly it was talking about this difference in mindset or vibrational level.

Where Is This Universe?

Don't go thinking that our linked universe is co-resident in our same space just because watchers appear and seem to disappear instantly from this heaven place. There has been "NO" credible evidence that our linked universe [and perhaps more than one] MUST be connected physically. The whole concept of quantum mechanics also gives us a better understanding that what we perceive as placement in time-space has little to do with reality. The time-space reality we enjoy is held together by common understanding of the environment we live in. Don't worry about everything you know today changing in a blink of the eye just because others decided not to think that way tomorrow. It doesn't work like that. Changing the group consciousness is not something that can be accomplished easily but there are some glimpses to it being done all the time.

- When someone lifts a car to save a child, it can't be done except in the modified consciousness area. There must have been a very strong yearning to have the slight change in reality and the creator also might have stepped in to allow for the change as well.

- When Peter walked on water, he certainly was able to

change his environment by changing his "life's resonance" but he also could have been joined by his fellow shipmates and don't discount Jesus, God incarnate, being around.

- When various monks were able to lift off the ground and fly, they had to stay in a state of meditation and prayer for some time before this little change in environment could be accomplished.

- Faith healing, or what has been termed that, surely shows a shift in our normal reality and a change in the common consciousness. Part of this is accomplished by convincing everyone that it can be done, convincing one's self that is has been done, and probably

- These guru characters that bury themselves for days without enough air, food, or warmth to assure survival, and they return unharmed must have shifted their consciousnesses.

- When Keely's experiments showed such promise and then fell apart as others tried them until he placed his hand on them, we can assume that something miraculous was happening. He, somehow, affected his life resonance and that affected the near-term reliability.

Life and the Out-wave

Just like matter has no end as Einstein and Wolff so eloquently described, life has the same characteristic. There is no specific end of consciousness. Some might say that this statement is the ravings of someone who simply cannot fathom an end of his or her life. Others would take a more religious approach and say that life ends and life in heaven or hell begins. Rather than trying to understand the oscillation characteristics of life's ebb and flow or the renewing of matter as it reaches the ends of our universe, one might better understand the concepts as symmetry existence rather than conservation of mass, energy and life. These things are more cyclic than constant.

Symmetry Not Conservation

As I mentioned slightly before, symmetry is more important and more realistic than conservation of universal elements. You may know about the theory of Conservation of Energy. Energy simply changes state rather than dissipates. It has an endless oscillation or vibration between Static and Kinetic elements. In a way this is a true statement and certainly we can look at our universe in this way and things seem to fit together. What if I were to tell you that static energy, or what we call static energy is a characteristic of the out-waves from our universe and kinetic energy is a characteristic of in-waves from an outside universe. We don't conserve energy. It is continuously replenished. As our universe outputs out-waves, they are converted to in-waves and returned.

The universe dual MUST run symmetrically. The only way to apply force in this universe is from an outside force. AND Guess what!!! All dimensional qualities of this universe MUST BE symmetric with its universe dual.

Oh, you are a smart one!!! You are thinking, "That's not symmetry, the examples I keep bringing up show the opposite to symmetry.

Got You With This One!!!!.

Remember that our linked universe has backward time. Therefore, decreasing matter backward in time is exactly expanding matter in their time. Both universes would experience expanding masses as matter is created over here, anti-matter is increased over there in reverse time.

Backward Living

Oh Boy! I've opened up a bag of worms now. If both universes sense time backwards to each other, there could never be interchange between universes------Right????-------

Wrong!!!!

In our linked universe, people experience time differently. It would be hard for us to understand how they see time, but I do think there is exchange between our universes so this oddness simply must be. I think that explanation will take one of those gurus on top of a mountain to explain. I'm having a hard time with forward time and I think I can still explain life without the extra confusion

Certainly there could be more universes, The Membrane [M-theory] suggests many more universes could be co-resident with our "visible one" and so do many of the string theories and the symmetry of our universe may be shared, but it could also be indelibly linked and presented in this work.

What I was saying is this. Matter, Energy, Life, Time and everything else in this universe stays in constant motion. Never increasing and never decreasing without a secondary outside force.

Let me give some examples-

- If **kinetic energy** decreases, it increases in our joined universe which in-turn causes our STATIC Energy to Increase and vice-versa. [If we make a conversion of energy in our universe, the opposite WILL occur in the joined universe.]

- If **Gravitational energy** decreases, it increases in our joined universe which in-turn causes the exact opposite to increase in our universe. This increase makes us have an apparent reduction in mass so we will have to look at it a little. [If we force a reduction in gravity, our joined universe WILL experience a reduction in gravity to compensate.]

- If **Photonic [Electro-magnetic] energy** increases, it increases in our joined universe which in-turn causes the exact opposite to decrease in our universe [With the opposite of light being something we call darkness, [Darkness WILL increase in the joined universe.]

- If **Internal Life energy** decreases, it increases in our joined universe which in-turn causes our External Life Energy to Increase and vice-versa. I know this whole external energy sounds weird, so we will look at that a little as well. [The main thing is that as internal life here increases, it WILL decrease over there.]

- As Time vector goes in one direction, it MUST go in the exact opposite direction on a joined universe. **[Time must go backwards. This, of course is from our point of view.]**

A general overview of this type of interactive universe duality is represented in the following tables.

Primary Universal Dimensional Comparison

Functional Characteristic	Particle Dynamo	Operational Dynamo	Ethereal Dynamo	Time Dynamo
Primary dimension	Fermionic	Electrical	Selfness	particulate time
Secondary Dimension	Gravitation	Magnetic	Spirituality	operational time
Tertiary Dim.	Nucleatic	Photonic	Soulness	Ethereal time
lowest dual	Fermio-Gravitation	Electro-magnetic	Self-Spiritness	Particulate awareness
Secondary Dual	Nucleo-Gravitation	Photo-magnetic	Self-soulness	Operational awareness
Highest order Dual	Nucleo-Fermonic	Photo-Electric	Spirit-soulness	Ethereal Awareness
Primary Time differential	Fermionic time	Electrical time	Egocentric time	x
2nd time differential	Gravitational time	Magnetic time	Spiritcentric time	x
Tertiary time differential	Nuclear time	Photonic Time	Soul-centric time	x
2nd derivative	particles	Forces	Life	Universe

What I've tried to do in this table is to express similarity of dimensional characterization between dynamos. Notice the union of the key characteristic and understand that everything is linked in some way. The universe would fall apart if it were not this way so we can interpret other dimensions by examining dimensions characteristics that we more easily can interpret. As life is given as emotional characterization, it is harder to sense than the physical elements of the fermio-gravitational [Structural] dynamo and

194

the force and stress relationships of the Electro-magnetic [Operational] dimensional dynamo.

Heaven Universal Dimensional Comparison

Functional Characteristic	anti-Particle Dynamo	anti-force Dynamo	anti-life Dynamo	inverse Time Dynamo
Primary dimension	inverse-nuclear	inverse-Photonic	Heavenli-ness	inverse Ethereal time
Secondary Dimension	inverse-particulate	inverse-Electric	Carnality	Inverse particle time
Tertiary Dimension	inverse-Gravitation	inverse-magnetic	Enlighteness	Inverse Op.-time
lowest dual	Nucleo-Fermonic	Photo-Electric	Enlightnened carnality	Ethereal Awareness
Secondary Dual	Fermio-Gravitation	Electro-magnetic	Heaven-Carnality	Particulate awareness
Highest order Dual	Nucleo-Gravation	Photo-magnetic	Heaven Enlighteness	Operational awareness
Primary Time differential	Nuclear time	Photonic Time	Soulcentric time	x
2nd time differential	Fermionic time	Electrical time	Egocentric time	x
Tertiary time differential	Gravitational time	Magnetic time	Spiritcentric time	x
2nd derivative	anti-matter particles	anti- Forces	second life	Heaven

This second chart describes the dual that is expected in our linked universe [I call heaven]. Please notice that some things appear to be direct duals of opposites while others change from electro-mechanical to particle and vice-verse. These variations have been described throughout this book and are collected here for convenience.

Certainly, these may not be the exact duals of our universe, but hopefully, it will give you an idea about how each one of the elemental parts connect with both internal dual

195

dimensions and external anti-dual dimensions.

To carry it one step further, we can establish that is the energy bond for the Particle dynamo is [Static and centripetal] forces and the operational dynamo is associated with Kinetic and centrifugal forces, the opposite would be true in the joined universe. The anti-particle dynamo would be associated with Kinetic energy and the anti-force Dynamo would be associated with Static energy in that world.

All this stuff is good information, but I know you really want to learn some more about how to attain a higher level of awareness or better life. As described in book 2 of the series, the ethereal dynamo is loaded with elements that can modify the universe and establish a person with a better understanding of and capability to operate in our universe better. If we are to understand consciousness, maybe a good point to review is our brain. While not very much of our true understanding comes from the twisted mass of nerves and memories, there are key elements of our brains that we saw in book 2. The main thing is that the brain patterns can be modified, controlled, and understood by understanding and modifying the brain vibrational baseline.

Brain Classification

SO many people base life on brain function, I think that we need to look at the thing for a minute and sort of expand on the "normal" definitions we have come to love. While the brain activity doesn't directly explain life, it may provide another level of insight. Let's investigate further. The brain functions are, many times, segregated into 5 parts; Epsilon, Delta, Beta, Theta, Alpha, and Gamma. While vibrational characterizations are not clearly understood, many try to test and experiment with the various brain frequencies just to see what happens.

Epsilon Brains

Epsilon or vibrations less than 1/2 hertz are not very normal, but there are some who can sort of suspend their brain functions and report **spiritual insight and ecstasy** as if it were an out of body experience.

Delta Brains

This is sometimes called the **consciousness of survival**. Delta or vibrations less than 1/2 to 4 hertz seem to be a **remedy for anger** and causes **confusion, disorientation, lucid dreaming, and a decreased awareness of the world**.

197

One might experience this brain function vibration level if he were in a **trance** or under hypnosis.

- **Medulla Oblongata-** As one would imagine, this brain component executes the most important functions of survival including regulating our life processes such as breathing, maintaining a steady heart rate and blood pressure, urination, and defecation.

- **Cerebellum** -Coordinates and controls voluntary movement, maintains balance and equilibrium while walking, swimming, riding, etc., stores memory for reflex motor acts, coordinates simultaneous subconscious actions, like eating while talking or listening etc.

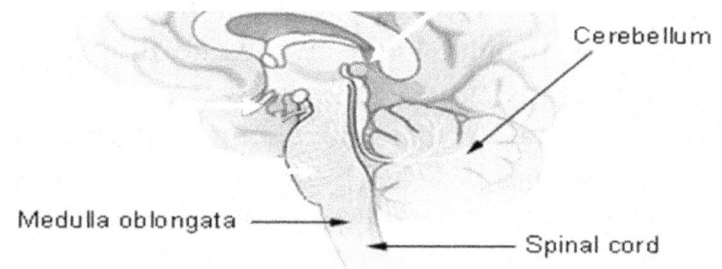

Theta Brains

This is sometimes called the **consciousness of sex**. Theta or brain function vibrations from 4 to 7 Hertz seems to cause **sexual arousal**. It also enhances **memory, focus, creativity and inspiration.** The section of the brain that becomes most active is the pons.

- **Limbic System-** Manages olfactory pathways and functions related to sex, rage, fear; emotions. The pineal gland shown below seems to be sensitive to much higher frequencies that the pituitary gland shown below.

198

- **Pons-** Manages Respiration. Has control over skin of face, tongue, teeth, expression, and level of arousal.

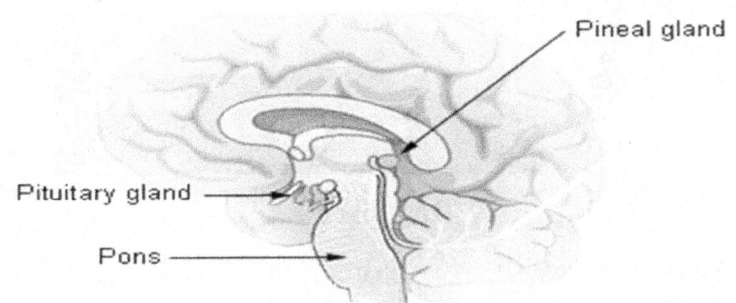

Alpha Brains

This level is sometimes called the **consciousness of self**. Alpha or brain function vibrations from 8 to 12 Hertz seem to **accelerate learning with reducing stress and elevating one's mood.** A person becomes more positive of his situation and may daydream and become relaxed without becoming drowsy. Particular areas of the brain seem to be excited by alpha waves including the following:

Cerebral Cortex- The outermost layer involved in the functions of learning, making decisions, and memory. The alpha and beta brains are established in the back portion. This contains the following:

- **Parietal Lobe-**Processes sensory input, sensory discrimination.

- **Occipital Lobe-**Concerned with visual interpretation.

- **Temporal Lobe-**Section that accommodates auditory performance, speech, and information retrieval.

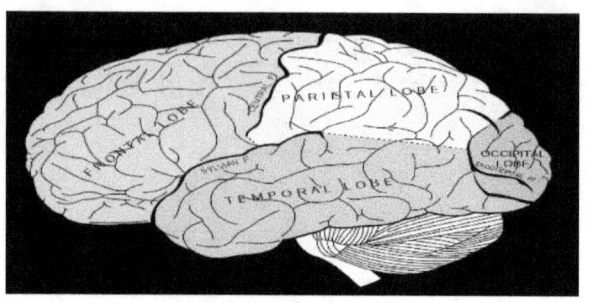

Beta Brains

Also associated with consciousness of self, Beta or brain function vibrations from 12 to 30 Hertz can be induced through active concentration and is associated with alertness, anxious thinking, and analytical problem solving. It also seems to increase judgment or decision making and **survival instinct**. It is as if we **processing information about the world** around us better in this state.

Gamma Brain

This is sometimes called the **consciousness of love**. Gamma or brain function vibrations from 30 to over 100 Hertz really starts us going right as our **cognitive skills and focus are heightened**. We have **increased compassion** and perception of reality. Additionally, we seem to take control over our bodies as **muscle develops, injuries begin to vanish and bones heal**.

Frontal Lobe- Slightly different than the other external lobes shown above, this one establishes memory and cognition. It also enables concentration, judgment, inhibition and personality development. Also associated with the gamma brain are the higher-level glands.

- **Thalamus** –Its main function is providing the brain information on what is happening outside the body.

200

- **Hypothalamus** –This section regulates emotions, hunger, thirst, libido and is responsible for maintaining the daily sleep and awake cycle. To show its strength, it also changes the pituitary gland which changes body homeostasis.

Above Gamma

Little is known above a gamma state except that it appears that the pineal gland becomes more active as vibrational ecstasy goes beyond normal gamma levels. We will look at the pineal gland a little more, but just like all the other vibrational dynamos the higher in vibration, the more kinetic the dimension becomes. The Etheric dimension works the same way. Note that all we have discovered to date are very low "Carnal" vibrations emanating from our brains.

Levels of Consciousness

The reason I brought the preceding section was to show that our brains are sensitive to vibration and that we are basically Carnal. As the frequency increases, the level of awareness or association with non-carnal life can be sensed. The reason is simple. Consciousness is a vibrational dimension and brainwave studies are not the only way to recognize the vibrational characteristics. A second way is something called chakra so let's look at some of these mystical things. One thing you will notice right away is that there are many levels not addressed in the brain function vibration frequencies normally studied.

According to the believers and the testing skeptics there are at least seven of these chakras or levels of consciousness.

- Root, Chakra--- "consciousness of food or survival"

- Sacral Chakra --- "Consciousness of Sex"

- Solar Plexus Chakra --- "Consciousness of Self"

- Heart Chakra --- "Consciousness of compassion and love"

- Throat, Chakra --- "Consciousness of the truth"

- Third Eye Chakra --- "Consciousness of our inner being"

- Crown, Chakra ---"Consciousness of the worlds beyond"

They are sort of represented by the diagram below.

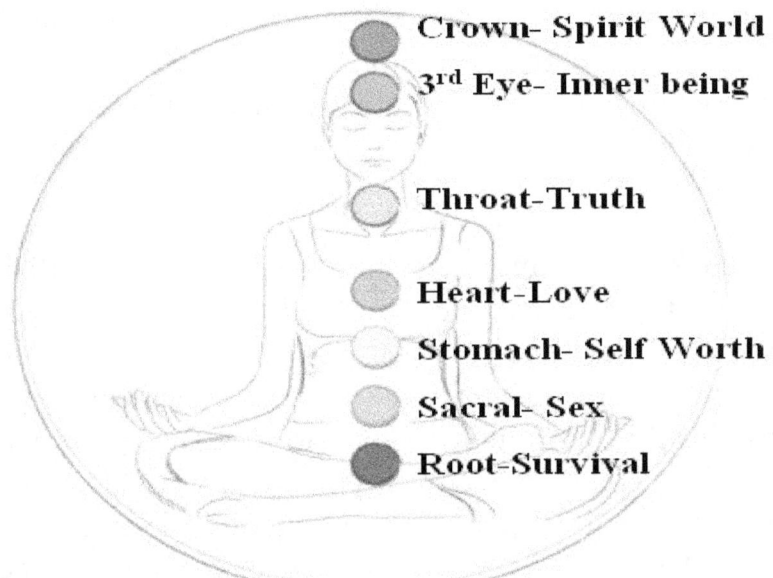

Crown- Spirit World

3rd Eye- Inner being

Throat-Truth

Heart-Love

Stomach- Self Worth

Sacral- Sex

Root-Survival

I know it sounds like I'm some guru from India talking about chakras, but it is a convenient way to discuss this dimensional component so I'm going to continue. I'm not putting on the towel on my head, but I may hum a little as I write this section.

Survival, Sex, Self

Whether we admit it or not, every one of us battles these things every day. The most basic root chakra is triggered when you get hungry and the sex one, well it shows up from time to time. If you can get past those you start considering self-worth and even love. Many people spend most of the time going back and forth between these two states. It's sort of like a yo-yo. Slow vibrations, higher vibrations, slower vibrations, higher vibrations and still higher vibrations when someone pays me a compliment or a figure out why a light bulb turns on in the refrigerator or I answer one of those "Are You Smarter Than a 5th Grader" questions. Answering one of the "Jeopardy" questions correctly might even get you into the heart chakra, but sex and self still will take control at some time.

Love and Truth

The heart chakra is a vibrational level associated with "Real" love rather than the sexual one. Sometimes this level happens naturally for a brief time and you can't seem to even think about yourself at all. If you work at it you can get to this level periodically throughout a day and look at people with true look. The Bible called it loving people as you love yourself. Anyway, most just think they get into love and it is more basic. That type of love puts you below the stomach again. Anyway, you must conquer love to some level before you can even get to a point that looks for "real truth" rather than "Vain truth" that we usually accept or desire.

Real truth is a truth that is truth no matter how it affects the event or who thinks it. It is usually not a popular truth or even the one you would hope for. It simply is. While it seems that this would be easy to understand and use. People almost never are tuned to this type of consciousness so they accept what they believe rather than what they should believe. Let's say you get an openness to understand real truth, there are still 2 more levels of consciousness to be considered. The next is called the third eye.

The Third Eye

The third eye is derived from the little gland in the brain I mentioned before called the pineal "pinecone" gland. The pineal has no apparent use, but it is thought to have been used by our brains at one time. After all, the gland didn't just grow there for no reason, so let's travel back to the Tower of Babel.

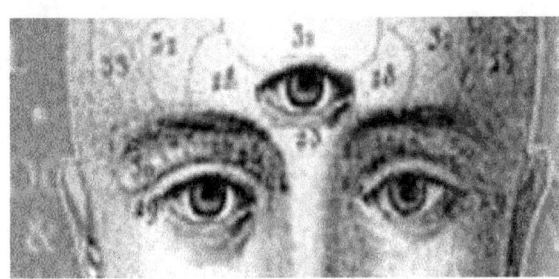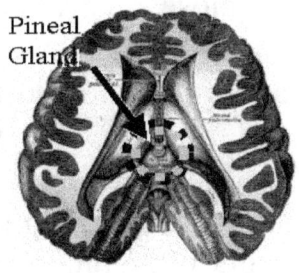

Pineal
Gland

Tower of Babel and Pineal

As I mentioned in book 2 of this series, the Tower of Babel is a huge structure that King Nimrod had built about 6 thousand years ago. A couple of things that may surprise you is that the Tower apparently was built in the country of Lebanon at a place called Baalbek rather than in Iraq as you have been told and the tower was associated with a huge world war when 1/3 of all the people on earth were killed. No one really knows what happened, but what can be derived from substantial evidence is that, all of a sudden, our brains lost most of their capability according to many ancient texts and how this brain loss was probably from some DNA modifying bacteria or something similar. I could also bring out the unusual fact that our current brain size is smaller than our earlier cousins, Neanderthal. While that fact is well known, what is not recognized is that this reduction in brain size shows that our brains began atrophying from disuse about 6 thousand years ago. I could bring out the fact that the entire

205

world was plunged into some type of Stone Age re-insurgence 5 to 6 thousand years ago and people seemed to become dumb as stumps for a while. The Biblical book of Jasher simply tells us that 1/3 of the people died, 1/3 of the people became like apes and 1/3 of the people were dispersed to places around the world because they could only talk to their close relatives. We can imagine that before this brain reducing started, we could do many things with our bigger brain we cannot do today. We can imagine that the pineal gland, prior to whatever happened 6 thousand years ago also was larger and might have been used by our ancestors. I could also bring up many other things that would make you wonder if the pineal gland used to allow us to do many things in the past, but I won't. Instead, let me tell you what this tiny, pea-shaped gland does.

- **Pineal Glands** in many non-mammalian vertebrates have a strong resemblance to the photoreceptor cells of the eye. Some evolutionary biologists believe that the pineal cells share a common were the ancestor to retina cells in the eye.

- **In some animals'** exposure to light of this gland can change the animal's biorhythm.

- **Some early vertebrate fossil skulls** have a pineal opening so that it probably had some vision characteristic.

- **The lamprey and the tuatara** both have this same type of pineal opening and this thing is photosensitive. The structures appear to include cornea, lens and retina,

- **The pineal gland is weird** in that it has profuse blood flow, second only to the kidney, so we can be sure that it once was of great importance. While doctors are perplexed at why this insignificant gland would need so much blood, it is obvious that whatever happened 6 thousand years ago made the extra blood flow unnecessary.

- Potentially this organ aided our capability to vibration our conspicuousness to higher levels. All these brain vibration levels are not exactly the same as consciousness vibration, because these are physical elements that stand in the particle dimensional dynamo, but there seems to be a correlation to the "Expected effect" to consciousness and the brain vibrations so we should not ignore the characteristics or the oddness of the pineal.

- **The brain of a 90-million-year-old bird** was found with a large parietal eye and pineal gland so it's been used for some time now to provide additional insight beyond normal seeing.

- **Production of melatonin** by the pineal gland is stimulated by darkness and inhibited by light. This melatonin stuff affects sex drive.

I know I brought this stuff up in the second book of the series but we are talking about vibrational oneness of our consciousness and there are some elements that must be repeated. We have a tiny organ that used to be huge and it used to be an aid in seeing, regulating moods and sex drive, but our bodies are still trying to supply it with enough blood to run a huge organ. Today the tiny little thing seems to have been abandoned by our bodies, but maybe we just can't see what it can do without vibrating a little. Vibrating allows us to understand our world.

Abraham Maslow

Anyway! This pineal gland/third eye was supposed to have given us the ability to understand the world around us. If we increase our vibrational level by unison with our environment "some call it meditation" or by other exotic means, we can sometimes get in tune with the world around us and here is the odd part. We can even affect it. Another way of saying this is that the 3rd eye thing is that "Self–Actualization" that Abraham Maslow talked about.

Positive Thinking

Somehow getting our vibrational levels in tune with the vibrational patterns of the elements around us allows us to be more intuitive. We can sense reactions needed to affect the environment. As we affect the environment we can change it. Now the changes are extremely subtle. You cannot, for instance cause money to fly off a tree, but you can somehow affect the conditions around you that will make it easier to accomplish particular tasks simply by concentrating on these tasks and believing that these things will be accomplished. I know it sounds like gobbly-gook. The problem is that the affect is demonstrated over and over and over again. Positive thinking and getting in tune with the vibrational pattern of the environment actually works. There is no doubt about it. The issue is trying to get into the level of consciousness needed to get the universe to "Bend" a little is not only hard, it also is not easily sustained once one gets to this level of consciousness.

As the vibrations increase, less and less of the conscious experiences relies on the carnal experience. Less and less of our "Life" is established in what we consider reality. Let's think of this whole consciousness a little deeper. Let me start over with a question.

Can your consciousness REALLY leave your body?

-

ABSOLUTELY!!!!!

I'm sure your first thought is that it can't and even after I discussed nears death experience and other similar accounts of people leaving this conscious world you still just can't get it in your head. Don't be so ready to close your mind to things that seem to be going on around us. It is becoming more and more apparent each year that astral projection, near death experiences, prophets seeing beyond this reality, and even reincarnations have been and are elements of the same characterization of the consciousness dimension. I know you think you are using your consciousness right now, but there is more to it that you would like to believe. Others simply shake their heads and believe that it might be sacrilegious to even suggest such a thing, Well! It is not!!! There is a reason so very many people have witnessed similar things. It is not because Satan enters one's body and forces the images into a consciousness. It is because that is how God made us. Let's take another look at the various well-known accounts of vibrating out of one's consciousness.

Near Death Consciousness

It is believed that over 10 million Americans have had Near Death Experiences and lived to tell about it and most concur with the general light tunnel and all of that.

Check out [http://www.nderf.org] and read the accounts. The picture is similar and seems to go along with a 12-dimensional vibrations to counter entropy universe.

- **Many say they feel a** "Whoosh" as they go through a tunnel much like the Bosh artwork depicted on the cover. The Whoosh is reported by deep mediators as almost a buzz as the vibration level of our being brings us to the brink of separation with our bodies. This is consistent with the 12-dimensional universe and the Ethereal vibrational pattern.

- **Many see an overwhelming brightness-** Of course no one actually sees light. They only see vibration and understand that there is a brightness. As one opens up their awareness of the things beyond the Carnal existence, what we interpret should go beyond brightness.

- **People say they FEEL the Light-** The warmth is oneness with the spiritual characteristic of the universe. I know a lot of this seems like some shaman waving his raccoon

210

claw over your head, but the idea here is that electro-magnetic interactions and ethereal interactions have a duality. Feeling warmth and light are when the consciousness go towards more "soul-like characteristics just like the magnetism and gravity components. Having life vibrations get slower toward entropy places us back to the low basics of those lower chakra things. When we go there, things get dark and colder. As one vibrates to the love levels, companionship gains warmth and a brightness. This goes beyond that characteristic.

- **People say they feel an intense love and feel totally at peace.** There is no doubt that love requires our consciousness to be removed from the lower vibrational levels so during these experiences there must be this loving feeling.

I know some people have had the opposite experiences, but that makes sense as well as vibrational levels can be forced slower to become more carnal if one really tries.

Out Of Body Consciousness

It is estimated that about 1/4 to 1/3 of the population depending on which study you look at have experienced SOME type of out of body thing. Sometimes it is just for an instant, a feeling that you already did something or know something that will happen, but many of us lose linkage with the COMMON consciousness we depend on. Here is what people say again.

- **Not Dreamlike**-They insist that the time they are away is not dreamlike, but instead it is close to reality. Don't for a minute believe that the spiritual world is not a reality. It is simply a different reality and for most of us it is very difficult to even glimpse this wonderful part of our dual universe.

- **Feel Powerful**-They usually sense power and freedom. Like I keep saying the way to gain these levels of insight is to increase one's vibrational level way beyond the entropy element of destruction and separation from God and the spirit world. That increase in vibration is what is ALWAYS associated with feelings of freedom from the rigorous nature of our "Collective consciousness.

- **Aware of the surroundings**-These people recognize and describe objects seen in these states with great accuracy.

Others, including many who initially were very skeptical, have verified this strange fact. Don't let people try to tell you that heaven or other universes are removed from the carnal one. They are co resident in space but separated by conscious characteristics with time so there is no doubt that objects could be sensed as objects.

Projection

One type of out-of-body experience is call astral projection. In this method of increasing one's conscious vibrations, people indicate that they sometime prepare several days and focus on a place that they wish to project to and they have various techniques to place themselves in a hypnotic state. Sometimes a simple word or "Mantra" is used to set up a situation. Once the initial conscious level is elevated, they feel like they are flying and they must fight not to fall into a deeper sleep level. And they feel at peace with the universe.

Additionally, we find that sometimes communication with others can be established in this state. It is not clear if the individuals are always real or self-generated, but there is a substantial amount of information that suggests that astral projectors can communicate with other people and sometimes these people are not alive.

A Common Thread

Hopefully, you are seeing that in almost all cases, people begin these experiences by blocking out the world including all feeling. Those who are forced in that condition by some tragedy don't seem to have any difference in this effect. They leave their bodies, get comfort or wisdom, sort of talk to comforting people, get a heightened sense of reality, can float, and when they get focused back on the "real world" they are plummeted back into it. Many times, these people are changed forever. I think this is something close to the crown chakra level that ancients attest to. The trip nearing the Crown chakra has changed them forever and I'll tell you why. Their consciousness has been vibrationally enhanced. It vibrates closer to the level needed to do this transfer thing, but in the mean time they become more aware of the feelings of others and become more self-actualized. Whether the "people they interact with are the cause of the vibrational enhancement or some other mechanism is at work, I do not know, but the entire life force of the person is enhanced. Many like it so much they go off and do it again if they can.

These astral projections and out of body episodes bring people closer to the vibrational level associated with the crown level, but one more thing is required to go all the way to this heaven place. Before I get to that, I think I had better figure out things about Carbon 12 "the building block of life" and the Anthropic Principle before we get to the death section.

Carbon-12 Impossibility

Today scientists can't even figure out how carbon-12 was made, much less many of the larger more complex atoms. Carbon-12 is made from Helium-4, and Beryilium-8. That seems straight forward except beryllium-8 is extremely unstable and lasts only 0.00000000000000001 or 10^{-17} seconds before it turns back into helium-4. It is not enough time to produce the combination. Even today, no explanation of how carbon-12 is made has had any real meat.

Carbon-12 seems to not be possible, but we still find Carbon-12 is--- building people.

It is as if people were here and existence went backwards in time from our existence. I know this still sounds odd to you because you are used to time going in one direction and that no one or nothing can go backward in time---especially not a universe. It would also mess up the next section of the book as death would be the moment of inception which I certainly am not getting into. What I need to shore up more is something I have been bringing out during this book, but I didn't exactly put a name to it. That something is called the Anthropic Principle because it will help us understand death more than Carbon.

The Anthropic Principle

The idea that consciousness affects reality is not new, but now it has a sophisticated name "The Anthropic Principle" or the Anthropic Universe Theory". I don't when it was named, but I missed it in the earlier books that rely on this very important factor in our universe. No one likes this idea because it makes the concept of time more difficult, but the whole Quantum Mechanics era has really destroyed our concept of stable time and stable reality, so, we are stuck with it no matter what. The whole "Power of Positive Thinking", "Think and Grow Rich" and all other concepts of the 70s which tried to convince you that how you consciously view reality will affect reality, are not only true, they affect your death as well. In the Anthropic World if you have faith of a grain of muster-seed, you can move a mountain, as Jesus said thousands of years ago and you can walk on water as demonstrated by Elijah, Elisha, Peter, and

215

Jesus did so many years ago. With the Anthropic Principle, science and religion can act as a single tool for us to understand God, the universe, and ourselves.

I know some of you are skeptical about your being able to shape reality with your thoughts and you don't see how that ability can affect your death so I will address these things in more detail in the next book of the series. If you were skeptical about life, I guarantee you will have a somewhat difficult time with the subject or death.

This Is Not The End

In the final book of this series, we will see that the "going to the light" and feeling the warmth and all the rest may not be the end. Nor is it a one-way tunnel. All the warmth stuff, friendliness, and other feelings seem to all be there, but death is a very strange component of life. Let me try to finish the study of life and light in a more open-minded way. While we are alive, let's also look at how easily the consciousness can leave the body [Alive or Dead]. Once the body becomes less significant, death may have less significance as well. There is a strong characteristic of humans making us become a three-component entity. If our Self, Soul, and Spirit all work together properly we gain true understanding of life and escape the horrors of death. If we don't allow them to work, our life experience is severely reduced and we have less and less impact on reality. That limiting effect leads to destruction, misery, and torturous death. The self, or carnal consciousness must be tamed by expanding our awareness and accepting the truths of life using our Spirit or non-carnal consciousness that starts with this self-actualization stuff. Finally, our soul must be understood and changed to allow transition into the universe of Heaven so that we can gain true happiness, be able to worship God directly, live a life after life with freedom, and eliminate our fears. As this is pretty much impossible alone, God incarnate promised to help us with his Holy Spirit to guide us in this adventure.

With that last little bit, I had better stop this book or I will begin preaching. The next book will expand these theories and truths to show how vibrational Resonance may play a very important part in our existence and the existence of our universe as we perceive it to be. Once we know about building the universe, we will conclude the series by gaining more insight into what death truly is.

Conclusions

Light, life, and death have similarities and differences.

- **Light is not Light**-We looked at how light was different than what has been imagined as light in the past. While the apparent effect of light can be initiated by placing stress on particles with electromagnetic [In-waves], the electromagnetic waves do not cause light. They cause electromagnetic fields. Light is established in our consciousness. To that end, the color red is completely different to all people. While we define it as the same, what might be a more vibrant red hue might be interpreted as less bright by another or even a completely different color by an animal seeing the same vibrations.

- **Electromagnetic vibrations**- don't light up anything. While they generally are present during feelings of light and visual comfort, sometimes, light can be sensed with our eyes closed or during near death experiences showing eyes are not the important part.

- **Sideways Light**- We discussed how light is transposed from normal life and if time were viewed sideways, the thing we call light would appear as a solid mass while life would simply have excursions where an entire lifetime could be viewed simultaneously.

- **Life Not DNA-** As we started studying life, it became apparent that life was not the same thing as DNA. Dead DNA and live DNA are similar in structure and forces on one versus the other would cause similar reaction.

- **Life was a different dimensional dynamo** from matter and forces on that matter. While it is different, nothing can be established in our universe unless life is instituted and a general acceptance by the combination of the conscious group must be assured before a true world can be generated. As people try to veer away from the common knowledge or viewpoint our world changes. When God stated that *"faith of a grain of mustard-seed would move mountains"* tells us how important our conscious minds are to our reality.

- **The Anthropic Universe Theory-** Tells us that each of us hold the world together. We cannot die or a piece of the universe will be lacking. Conservation of Energy won't allow it.

- **Life Between Lives-** The old Purgatory looks a little different than described in many Church Dogma. Instead, substantial research has shown that people must learn more than they typically learn about the non-carnal existence so they reenter life several times.

- **God Exists-** Unfortunately, the universe cannot exist alone. A controller keeps everything from slipping back into entropy and disappearing. This God is the creator of everything and from the other books in the series, I hope that you understood how very important it is for the God stability to work in this universe. In-waves and out-waves stabilize the non-living. God stabilizes the living/conscious elements of the universe.

- **God is Light**- In the form of the Holy Spirit or Holy Ghost, we found confirmation that this type of light was extremely important in not going to what we call Hell.

With that I must end this book.

About the Author

Steve Preston is a long lime author of scientific, esoteric facts. His books focus on the painful truths rather than whitewashed details that make us comfortable. If you are interested in the truth instead of comfort, please review other works by Mr. Preston as shown below. The images are some from Egypt taking the older version of a taxi. To the right the writer is shown in the Jewish Negev desert of Israel where the Dead Sea Scrolls were found.

To the left below are a couple of pictures as we searched the New Zealand caves searching for ancient Maori artifacts and the last image is of the author investigating the Acropolis concerning ancient Athens Greece.

Ancient Technology
Titan Gods- History of the Ancient Giant/gods
Kingdoms Before the Flood- Pleistocene humans
Victory of the Earth- History of our Earth
Amazing Technology- Descriptions of prehistoric capabilities
Ties to Planets and Space
Where UFOs Go- Discussion about UFO pilots
Not from Space- UFOs are not from space.
Martians- Ancient Life on Mars
Living on Venus- Venus before the Pleistocene Extinction
Ancient History of Flying- Ancient flying
When Giants Ruled the Earth- History of the Titan Giants
Ties to Egypt
Moses Saved Egypt- How the Jews eliminated the Hyksos
Mysteries of the Exodus- Proofs of the Exodus
Scythians Conquer Ireland- A History of Ireland
Truth About Phoenicia- The Evidence -First in America
Egyptian Foreigners- Story of the Amalekite invaders
Mysterious Pyramids- Who made the Pyramids?
Secrets of Thoth- Details of Emerald Tablets
Truth About the Hyksos Pharaohs- Horrors of the Hyksos
War
Behind the Tower of Babel- Story of the Bharata War
Driven Underground- Fear in the Bharata War
Four Armageddons- The 4 major wars that destroyed mankind
Six Deaths of Man- Destructions of mankind
World War Before- The Pleistocene War
World War with Heaven- The Angel and Anak War
World War Zero-The Bharata War
Sex Crazed Angels- What caused the Heaven War?
Ties to America
Who Really Discovered the Americas? the Phoenicians
Romans found America- Discovery of Copper
Mysterious PreIncan Journey- Technology of the Inca

New Look at Biology and Self

Understand Your Heart- New Discoveries of the Heart-brain
DNA of Our Ancestors- Tracing DNA of ancient man
God Didn't Make The Ape- New science on ape Evolution
Lizard People- Mutated People of the Bharata War
Creation and Death of Dinosaurs- Why Dinosaurs died
Races of Men- Tracing DNA of Humans
Tracing Cro-Magnon to Jesus- Follow new findings
Self, Soul, Spirit- Three components of Life
Self-Virtualization- New science of reality
True Happiness- Self Actualism and Beyond
Life Resonance- Unusual capabilities of men
Awaken the Departed- We can talk to the Dead
Biophotonics and Healing- How Photonics used in medicine
Homo-Erectus as a Man- Characteristics of Homo-Erectus types

www.ingramcontent.com/pod-product-compliance
Lightning Source LLC
Chambersburg PA
CBHW051642170526
45167CB00001B/292